SCIENTIFIC BASES FOR THE PRESERVATION OF THE HAWAIIAN CROW

Committee on the Scientific Bases for the Preservation of the Hawaiian Crow
Board on Biology
Commission on Life Sciences
National Research Council

NATIONAL ACADEMY PRESS
Washington, D.C. 1992

National Academy Press • 2101 Constitution Avenue, N.W. • Washington, D.C. 20418

NOTICE: The project that is the subject of this report was approved by the Governing Board of the National Research Council, whose members are drawn from the councils of the National Academy of Sciences, the National Academy of Engineering, and the Institute of Medicine. The members of the committee responsible for the report were chosen for their special competences and with regard for appropriate balance.

This report has been reviewed by a group other than the authors according to procedures approved by a Report Review Committee consisting of members of the National Academy of Sciences, the National Academy of Engineering, and the Institute of Medicine.

This study by the Board on Biology was supported by the Department of the Interior, U.S. Fish and Wildlife Service under grant number 14-16-0001-91578.

Any opinions, findings, conclusions, or recommendations expressed in this publication are those of the authors and do not necessarily reflect the views of the U.S. Fish and Wildlife Service.

Library of Congress Catalog Card No. 92-60711
International Standard Book Number 0-309-04775-7

Additional copies of this report are available from:

National Academy Press
2101 Constitution Avenue, N.W.
Washington, D.C. 20418

S-607

Printed in the United States of America

COMMITTEE ON THE SCIENTIFIC BASES FOR THE PRESERVATION OF THE HAWAIIAN CROW

W. Donald Duckworth *(Chairman)*, President and Director, Bishop Museum, Honolulu, HI
Tom J. Cade, The Peregrine Fund, Boise, ID
Hampton L. Carson, University of Hawai'i, Honolulu, HI
Scott Derrickson, National Zoological Park, Front Royal, VA
John Fitzpatrick, Archbold Biological Station, Lake Placid, FL
Frances C. James, Florida State University, Tallahassee, FL

Special Advisors

Cynthia Kuehler, Zoological Society of San Diego, San Diego, CA
Stuart Pimm, University of Tennessee, Knoxville, TN

National Research Council Staff

Donna M. Gerardi, Study Director
Norman Grossblatt, Editor
Scott Olson, Project Assistant
Mary Kay Porter, Project Assistant

COMMISSION ON LIFE SCIENCES

National Research Council Staff

The National Academy of Sciences is a private, nonprofit, self-perpetuating society of distinguished scholars engaged in scientific and engineering research, dedicated to the furtherance of science and technology and to their use for the general welfare. Upon the authority of the charter granted to it by the Congress in 1863, the Academy has a mandate that requires it to advise the federal government on scientific and technical matters. Dr. Frank Press is president of the National Academy of Sciences.

The National Academy of Engineering was established in 1964, under the charter of the National Academy of Sciences, as a parallel organization of outstanding engineers. It is autonomous in its administration and in the selection of its members, sharing with the National Academy of Sciences the responsibility for advising the federal government. The National Academy of Engineering also sponsors engineering programs aimed at meeting national needs, encourages education and research, and recognizes the superior achievements of engineers. Dr. Robert M. White is president of the National Academy of Engineering.

The Institute of Medicine was established in 1970 by the National Academy of Sciences to secure the services of eminent members of appropriate professions in the examination of policy matters pertaining to the health of the public. The Institute acts under the responsibility given to the National Academy of Sciences by its congressional charter to be an advisor to the federal government and, upon its own initiative, to identify issues of medical care, research and education. Dr. Kenneth Shine is president of the Institute of Medicine.

The National Research Council was organized by the National Academy of Sciences in 1916 to associate the broad community of science and technology with the Academy's purposes of furthering knowledge and of advising the federal government. Functioning in accordance with general policies determined by the Academy, the Council has become the principal operating agency of both the National Academy of Sciences and the National Academy of Engineering in providing services to the government, the public, and the scientific and engineering communities. The Council is administered jointly by both Academies and the Institute of Medicine. Dr. Frank Press and Dr. Robert M. White are chairman and vice chairman, respectively, of the National Research Council.

PREFACE

Much has been written about the unique and fragile environment of the state of Hawai'i. Formed by up-welling lava from the ocean floor, the Hawaiian archipelago is the world's most isolated island group, located some 4,000 km from the nearest continental land mass and 3,000 km north of the Marquesas Islands. The islands of Hawai'i range from sea level to over 4,000 m in elevation and receive an annual precipitation ranging from 200 mm to over 10,000 mm. Together, the isolation and diverse physiography of the islands influenced the evolution of a unique biota characterized by low phylogenetic diversity, high endemism, and spectacular adaptive radiations. Many groups of organisms--such as the silverswords (Asteraceae), lobeliads (Lobeliaceae), land snails (Achatinellidae, Amastridae, and others), pomace flies (Drosophilidae), and Hawaiian honeycreepers (Drepanidinae)--radiated from single ancestral populations into diverse assemblages of closely related species occupying a broad range of habitats.

The extraordinary array of unique life forms on the Hawaiian archipelago that evolved over millions of years has declined at an ever-increasing rate since the arrival of humans and their intentional and accidental fellow travelers--exotic plants and animals--and today Hawai'i has the distinction of being the extinction capital of the world. Relatively intact native ecosystems have generally survived only at higher elevations, and many extant species now occupy only small remnant portions of their former ranges. Hawai'i consists of only 0.02% of the land area of the United States, but it has sustained nearly 75% of the nation's documented plant and animal extinctions, and native Hawaiian species are a substantial proportion of the species included on the U.S. list of *Endangered and Threatened Wildlife and Plants*[1]

Against that tapestry of severe habitat loss and loss of biological diversity, the efforts to save a single species, such as the Hawaiian Crow (*Corvus hawaiiensis*), or 'Alala, might seem to have little importance. However, single-species conservation efforts can be justified on moral, ethical, legal, economic, and scientific grounds, and it is especially important to recognize that programs to save single "indicator species" can provide a foundation for broader conservation and education initiatives. Research, habitat restoration, and educational efforts undertaken for the recovery of a single endangered species clearly can benefit a wide range of species and habitats if properly designed and effectively implemented.

---

[1] U. S. Fish and Wildlife Service. 1991. Endangered and Threatened Wildlife and Plants. Title 50 17.11 and 17.12, July 15, 1991. Washington, D.C. : U.S. Fish and Wildlife Service, 37 pp.

In response to a request from the U.S. Fish and Wildlife Service, the National Research Council's Board on Biology established the Committee on the Scientific Bases for the Preservation of the Hawaiian Crow ('Alala) in September 1991. Its task was to review the existing data pertaining to the 'Alala while focusing on several scientific issues:

- "Assess to the extent possible the causes of population trends of the 'Alala in the wild;
- "Assess to the extent possible the causes of population trends of the 'Alala in the captive population;
- "Evaluate options for action to maintain or increase numbers of the 'Alala in both the captive and wild populations;
- "Estimate the minimum viable population for survival of this species; and
- "Determine the advisability of adding genetic stock from the wild flock to the captive flock. If it is advisable, the 'form' in which this genetic material should be taken will be specified: e.g., eggs, fledglings, juvenile birds, or adults."

Current options for recovery of the 'Alala vary; they range from bringing all birds into a captive propagation program to leaving the wild population undisturbed. From the assembled data and their evaluation, the committee has developed a set of recommendations designed to assist interested parties in working effectively to aid the recovery of the 'Alala.

The committee held three meetings: two in Hawai'i (Honolulu, October 1991; Captain Cook, January 1992) and one at the Arnold and Mabel Beckman Center, Irvine, California (December, 1991). In addition, a subcommittee meeting was held at the Beckman Center (February 24-25, 1992). The committee collected and reviewed data on the 'Alala from biologists in federal and state agencies, other scientists and individuals. It visited the Olinda captive-breeding facility on the island of Maui. We are especially grateful to the owners and personnel of the McCandless Ranch on the island of Hawai'i for hosting the committee on a 2-day fact-finding trip on their property. The opportunity for the committee members in attendance to observe the 'Alala flock in nature was invaluable to the development of this report.

Many persons assisted the committee in preparing this report, and we are grateful to them. The committee is especially indebted to those who submitted information for its review:

Carolee Caffrey, University of California, Los Angeles
Robert Fleischer, National Zoological Park
Patricia Rabenold, Ohio State University
Cheryl Tarr, Pennsylvania State University

Special thanks are also due to the following persons, who were guests at committee meetings and participated in our discussions:

In Honolulu
Carter Atkinson, USFWS
Michael Buck, Hawai'i DLNR
Paul Conry, Hawai'i DLNR
Reggie David, Hawai'i Audubon Society
John Engbring, USFWS
Jon Giffin, Hawai'i DLNR
James Jacobi, USFWS
Dana Kokubun, National Audubon Society
Barbara Lee, Alala Foundation
Robert Pyle, Bishop Museum and Hawai'i Audubon Society
Karen Rosa, USFWS
Nohea Santimer, Alala Foundation
Charles Shockey, U.S. Department of Justice
Peter Simmons, McCandless Properties
Robert Smith, USFWS
Paul Spaulding, III, Sierra Club Legal Defense Fund
Carole Terry, Hawai'i DLNR
Ronald Walker, Hawai'i DLNR
Moani Zablan, Alala Foundation

On Maui, Olinda Captive Breeding Facility
Renate Gassman-Duvall, former contract veterinarian with Hawai'i DNLR
Fern Duvall, Hawai'i DNLR
Judy Pangelinan Subaitis, Hawai'i DNLR
Wayne Taka, Hawai'i DNLR

On Hawai'i, Captain Cook and McCandless Ranch
Paul Banko, USFWS
Andrew Berger, University of Hawai'i
Paul Breese, Honolulu Zoo
John Phillips, Professor Emeritus, Stanford University
Cynthia Salley, McCandless Ranch
Nicholas Santimer, McCandless Ranch
Nohea Santimer, McCandless Ranch
Peter Simmons, McCandless Ranch

Throughout the course of its deliberations, the committee was careful to maintain a steady focus on the scientific evidence available concerning the 'Alala and the implications of that evidence in relation to present and future efforts for recovery. Aware of past controversy and current disputes over recovery plans and techniques, which are the responsibilities of state and federal agencies and individuals, the committee assiduously adhered to its primary task. In this regard, the committee is grateful for the understanding and support of all parties associated with the study. The value of this approach, coupled with the considerable talent and experience of the committee members, is evidenced by the information and results described in the report.

Finally, the committee wishes to acknowledge with gratitude and admiration the vital role played by the National Research Council staff in all aspects of the project. The dedicated professionalism of NRC study director Donna Gerardi facilitated and expedited the committee's activities throughout the study. In addition, Alvin Lazen and Norman Grossblatt added considerable experience and effort to the endeavor.

W. Donald Duckworth, *Chairman*
Committee on the Scientific Bases for the Preservation of the Hawaiian Crow
May 1992

CONTENTS

EXECUTIVE SUMMARY

The Hawaiian Crow (*Corvus hawaiiensis*), or ‘Alala, once an inhabitant of large forested areas of the island of Hawai‘i, is now found only in the wild in a relatively small area of the central Kona coast, specifically on the privately-owned McCandless Ranch. The decline of the ‘Alala is part of a larger phenomenon of reduction and extinction of forest birds throughout Polynesia that has been associated with human colonization. Thus, its decline is not an isolated ecological event, but rather a symptom of underlying ecological problems. Of particular concern here is the ecology of montane forests on the western slopes of the island of Hawai‘i, known as the Kona district. In other habitats along the Kona coast where the ‘Alala has already disappeared, numerous endemic bird populations have also become extinct or are endangered and rapidly declining. The ‘Alala is the most conspicuous member of this group and can be viewed as a classic indicator species.

Three bird species that were known only on the Kona slopes are already extinct: the Greater Koa Finch (*Rhodacanthis palmeri*), the Lesser Koa Finch (*R. flaviceps*), and the Kona Grosbeak (*Chloridops kona*). They were seed-eating birds and fed primarily on koa and other seeds, on hard, dry fruit and seeds of the naio (*Mycoporum sandwicense*), and on lepidopteran larvae (Scott et al., 1986). All three were extinct by around 1900. It is reasonable to assume that the ‘Alala's precipitous decline has occurred because the problems that caused the prior extinctions have not gone away. The systemic problems, and not just the decline of the ‘Alala, must become the focus of scientific and conservation attention. As long as the ‘Alala exists in the wild, it can help researchers to identify facets of the systemic problems at the ecosystem level, and perhaps provide insight on how managers can remedy them and prevent additional extinctions on the Kona slopes.

The Wild Population of ‘Alala

The ‘Alala is an omnivorous, but primarily fruit-eating, forest-inhabiting corvid. Both in its reliance on fruit and in its restriction to a forest habitat, the ‘Alala differs from the widespread and familiar crows of the continents. The ‘Alala is far more specialized.

As of April 1992, it is known that at least 11 ‘Alala exist in the wild in at least three and possibly as many as five breeding territories on the McCandless Ranch (J. Engbring, pers. comm., 1992). In spite of the precipitous decline elsewhere, numbers of ‘Alala observed on the McCandless Ranch appear not to have changed drastically since 1976. The existence of other small populations on adjacent lands cannot be ruled out.

The ‘Alala used to be found in relatively low densities from the Ka‘u District in southeastern Hawai‘i to North Kona, north of Hualalai. Although extensive field studies have been conducted in those areas, many aspects of the natural history and behavior of the ‘Alala are still poorly understood. Such information remains critical to developing and implementing appropriate preservation efforts. More information about the habitat is also necessary to develop efforts to preserve other native species.

A single cause for the decline in the wild ‘Alala population cannot be identified; however, three categories of factors have been directly and indirectly implicated by previous studies: habitat and food, exotic predators, and exotic diseases and parasites. Agricultural development by the Polynesians destroyed most of the original dry lowland forests, restricting the ‘Alala to largely wet, open, mid-elevation ‘ohi‘a (*Metrosideros polymorpha*) and koa (*Acacia koa*) forests. In the post-European settlement period, extensive ranching, logging, and the practice of allowing feral ungulates--primarily cattle (*Bos taurus*), but also pigs (*Sus scrofa*), sheep (*Ovis aries*), and goats (*Capra hircus*)--to roam in the forests have degraded the forests, reducing the native understory of fruit-bearing trees and shrubs. Even the current habitat of the remaining ‘Alala is not necessarily prime habitat for the species. The extent of collecting and killing by humans is difficult to know, but it has probably been important, even in recent decades, and might have contributed to the high adult mortality of recent years off the McCandless Ranch. Introduced black (*Rattus rattus*) and Polynesian roof (*R. exulans*) rats prey on eggs and nestlings, and introduced mongooses (*Herpestes auropunctatus*) and cats (*Felis catus*) undoubtedly prey on fledglings on the ground. Two introduced diseases that are widespread among native Hawaiian birds have been shown to affect the crow: avian malaria and avian pox.

The committee's analysis of data obtained from census efforts during the 1970s and 1980s on a small sample of banded ‘Alala reveals that during that period of precipitous decline clutch size of the ‘Alala was somewhat smaller than temperate *Corvus* species, but sample sizes are small and comparative data for other tropical insular species are lacking. On the McCandless Ranch, the ‘Alala continued to produce fledglings at rates somewhat lower than those of temperate corvids, and fledging rates might even have been slightly lower elsewhere on the Kona coast. Juvenile (0-1 year) survival in the Kona District was comparable with that of other corvids that are not endangered. Except on the McCandless Ranch in the last 14 years, however, death rates of adult ‘Alala in the wild were inordinately high. We do not know why adult survivorship has been lower than would be expected for a *Corvus* species or why it has been higher on the McCandless Ranch than at other sites.

Avoiding Extinction

What follows is a summary of the committee's major findings and recommendations. Each is examined in further detail in Chapter 7.

EXECUTIVE SUMMARY

The federal Alala Recovery Plan (Burr et al., 1982) and the state Alala Restoration Plan (Burr, 1984) are admirable documents that contain concise summaries of the history and status of the species and sound recommendations for recovery. Each places priority on the protection and restoration of native habitat, the study of disease and predator control, and recommends the management of the captive and wild populations. The federal plan designates critical habitat that needs protection, and the state plan sets priorities for how that land should be incorporated into a conservation scheme. In the decade since the federal plan was issued, however, it has not been implemented, and the ʻAlala has declined further.

A reliable estimate of the number of ʻAlala that would constitute a minimum viable population for long-term survival of the species is not now possible. Given the small size of the current population, the species will be in danger of extinction for a long time. We conclude that there is a high probability that the current population in the vicinity of the McCandless Ranch will become extinct as a result of chance events in 1-2 decades unless its numbers and geographic range are increased. The committee is unanimous in its opinion that only an active management program can prevent the wild population from becoming extinct.

We do not recommend that all the birds should be brought into captivity. A viable population of ʻAlala in the wild should be maintained and increased. To avoid extinction, the sizes of both the wild and captive populations must be increased to provide demographic and genetic security. When populations are as small as these, demographic accidents and random environmental disturbances are likely to cause extinction. Considering the small size of the remnant populations, the committee recommends joint management of the wild and captive populations as a single unit. Joint management will require that the identity of all existing birds be known. The possibility that additional ʻAlala survive on both public and private lands in Hualalai, Honaunau, and the Kaʻu District needs to be thoroughly investigated by the U.S. Fish and Wildlife Service and the state of Hawaiʻi, because additional birds would be crucial to recovery of the species. Any additional birds would both increase the existing gene pool and decrease the probability of demographic accidents.

The goal of joint management of the captive and wild populations is to increase the density and distribution of the population as rapidly as possible. To accomplish the joint management of captive and wild populations, a new recovery team or advisory working group for the ʻAlala should be established that includes state and federal biologists; other professional biologists who are experts in avian ecology, captive propagation, reintroduction, long-term population biology of birds, and population genetics; an avian veterinary pathologist; an aviculturist; and a representative of the private sector, preferably a private landowner or land manager. This combination of experts will provide the knowledge and skills necessary for joint management of the wild and captive populations. The recovery team must work together continually to plan, coordinate, and implement all aspects of the recovery plan. A long-term, well-funded arrangement must be established.

For recovery and sustainability of the 'Alala, as for other endangered species in Hawai'i and elsewhere, habitat maintenance and restoration are essential. Unless the causative factors of decline are identified and corrected, preservation efforts for the 'Alala will be compromised. Through habitat preservation and holistic management of cattle ranches, and control of predators, some potential causative factors can be mitigated.

The committee strongly urges the state of Hawai'i and the U.S. Fish and Wildlife Service to establish at least one major forest preserve along the Kona coast. In addition, cattle ranches should be managed in ways that provide critical habitat for native flora and fauna. The habitat of the wild population of 'Alala should be managed aggressively to control predators--including mongooses and rats--and the impacts of exotic plants and feral ungulates.

Techniques must be developed for managing the wild population without risk of injury to the birds. Increasing their numbers in the wild will require knowing what they need to survive, remain healthy and reproduce successfully. Most of the information needed to support the management of the 'Alala as its numbers increase can be gathered noninvasively, with minimal disturbance of the wild birds. Some basic biological facts about their social behavior are still unknown. Most urgently needed are data on habitat requirements, variations in food resources available and required through the seasons, foraging behavior, physiology and disease, social behavior, and demography. Those data will provide specific direction for management of the property that supports the final wild population as 'Alala numbers increase and of additional forest preserves after reintroduction or recolonization. Short-term measures that can be started immediately include improved forest management, control of predators, supplementation of the 'Alala's food supply, study of disease in introduced species of birds that are living in the same habitat, and a public-education program.

Long-term measures that require additional research, planning, and the cooperation and coordination by a recovery team include development of a safe and effective vaccination program for avian pox and control of other diseases and reintroduction of young hatched in captivity. The latter actions will require careful analysis of methods used for the reintroduction and translocation of other endangered species.

The Captive Population of 'Alala

Through research and experimentation biologists have developed manipulative techniques that have been highly successful in the preservation of endangered species. In the case of the Peregrine Falcon (*Falco peregrinus*), for example, captive breeding and reintroduction enabled the restoration of breeding populations in large portions of its former range at a rate far exceeding that which would have occurred through natural colonization. Those techniques have usually been applied to complement conventional field research and management efforts to preserve habitat and restore threatened populations.

EXECUTIVE SUMMARY

Owing to the small number of 'Alala now remaining, complete recovery might seem hopeless; however, similar cases around the world that have used intervention measures have had striking success and offer hope that coordinated recovery efforts can be successful. To date, 39 species of corvids, including 9 *Corvus* species, have been bred successfully in captivity.

The management and husbandry of the captive 'Alala at Olinda captive breeding facility are inadequate, and there is room for modification and improvement of existing methods and procedures. The Olinda facility should be expanded. Management of the entire population at one site, however, is risky. The committee concludes that a single propagation facility is not adequate. A second facility should be built that provides an optimal environment for successful breeding of the 'Alala. An optimal facility would have additional crow enclosures that enable the safe capture and manipulation of the captive birds and incorporate the designs of experienced aviculturists from other captive-propagation programs. The general characteristics of an optimal facility (or facilities) would include institutionalization of routine and emergency veterinary care, adequate on-site veterinary facilities, a consistent pathology program, and an up-to-date library with access to journals and information at zoos on the mainland. Continual review of the avicultural literature and communication with colleagues are essential for developing or applying new techniques and procedures. A second facility should be established on the Hawaiian islands to minimize the potential for loss of the captive population to a disease outbreak or other catastrophe.

Other actions require extensive planning, cooperation, and coordination. For example, state and federal agencies should seek advice from outside experts about the design of an optimum captive habitat for the 'Alala. Consideration should be given to developing a more "natural" captive environment, constructing lower and wider breeding enclosures to facilitate capture while minimizing the risk of injury or trauma, providing breeding enclosures with multiple compartments to allow for the quick isolation and capture of males, provisioning each breeding enclosure with multiple nest baskets to stimulate nesting, and increasing the size of juvenile enclosures to promote improved flight conditioning. Staff training will be required, and protocols for standard and nonstandard procedures will have to be developed, circulated widely for peer review, and then implemented.

Husbandry in the captive-breeding program must be improved. Both short- and long-term actions are recommended. Some actions that can be instituted immediately are re-mating birds that have been paired unsuccessfully for more than 2 years, modifying enclosures to facilitate capture of and removal of males after the first egg of each clutch is laid, allowing females to incubate eggs 5-7 days before removal for artificial incubation, monitoring daily food intakes to assess dietary components and overall nutrition, and feeding the birds a more frugivorous, low-acid, low-iron diet.

Well-trained personnel are also essential. To remedy current shortcomings, a full-time director should be hired to serve as both a program administrator and curator. The person should be knowledgeable about aviculture and ornithology and be up to date on avicultural techniques, and would be responsible for the overall training of keeper staff, time management, facilities administration, public relations, and fund-raising. A full-time on-site avian veterinarian, a full-time on-site aviculturist, and at least two additional animal keepers are also needed. Equally important is the establishment of a long-term avicultural training program for staff.

In addition to the establishment of more than one captive-breeding facility, longer-term actions include the development of techniques for semen collection and artificial insemination, expansion of the captive flock until there are at least 40 productive pairs, and development of chick-rearing protocols that maximize proper socialization and independence. Those actions will require additional funding, training, planning, cooperation and coordination.

Modern equipment is needed if the captive-breeding program is to be successful. The program should have a full veterinary laboratory and clinic (equivalent to the facilities of a small-animal practice) that includes quarantine areas, an x-ray machine, updated video-monitoring equipment, a library, and a hatching room and brooding facility. It also needs access to a pathology laboratory on the Hawaiian islands, protocols for pathological analyses that can be performed on site, and long-term arrangements with a board-certified pathologist or pathology center for pathological studies.

Genetic Considerations

Although the historical decline of the 'Alala must have been accompanied by a loss of genetic variation, it is not clear whether the extant wild population is suffering from inbreeding depression caused by the loss of genetic variation. Recent DNA analysis shows the captive flock to be moderately to highly inbred. The 'Alala population on the McCandless Ranch is so close to the wild sources of some of the captive birds, however, the addition of wild adult birds from the ranch to the captive population for genetic reasons alone would be expected to provide no more than a very minor measure of new genetic variation to the captive population, and their removal from the wild would adversely impact on the extant wild population's potential productivity. Any potential beneficial genetic augmentation of the captive flock should be accomplished by removing eggs from the wild and hatching them in captivity.

Even extant genetic variation can be retained only if the population is increased rapidly. New genetic studies of the wild 'Alala are not recommended over the short term because the numbers are so small and the amount of information gained would not affect the principal recovery or habitat management actions that need to be implemented now. The capture of additional adult birds for captive breeding for genetic reasons alone should have a very low priority, because the potential genetic benefits would probably be minimal, and because wild-caught adults would have a reduced likelihood of breeding. Similarly, the release of captive birds on the McCandless Ranch solely for the purpose of augmenting genetic variation is not advised. Any releases must be part of a full-scale management plan that has demographic augmentation as its primary goal.

Options for Management of the ʻAlala

The committee's report suggests several options for the management of the ʻAlala. We give the highest priority to the removal of first-clutch eggs from the wild population beginning in the 1993 breeding season. Egg removal ("egg-pulling," "double clutching," "multiple clutching") has become a successful method for augmenting the natural reproductive output of wild birds, and has been used successfully with other endangered species. It relies on the capability of most female birds to renest after clutch removal or to continue laying eggs beyond the normal clutch size if eggs are removed in sequence as laid. On the basis of what is known about other corvids and birds generally, the failure to renest after removal of the first clutch will probably not be more than an infrequent occurrence or a rare peculiarity of particular females.

We give the egg-removal option a high priority because it allows for simultaneous augmentation of the captive and wild populations without removing adult birds from the wild or seriously compromising the wild population's social organization or productivity. The use of egg removal and later reintroduction of young into the wild is obviously a long-term strategy and requires an active recovery team and optimum incubation and chick-rearing facilities on or near the McCandless Ranch. Results should be evaluated annually so as to make recommendations about what to do with the crows that are produced in captivity.

The great advantage of egg removal is that it provides a way to take substantial numbers of eggs from the wild for hatching and rearing of young without depriving the wild birds of opportunity for natural reproductive output in the wild. The combination of natural and artificial hatching and rearing yields many more young than could be produced by the unaided wild population. The young reared in captivity can be used in several ways: they can be released back into the wild, they can be retained in captivity to augment the captive-breeding stock, and they can be used to learn more about the biology of the species.

Optimum conditions for artificial incubation have not yet been worked out for the ʻAlala, but this deficiency can be partly offset by ensuring that the wild eggs receive some natural incubation (about 5-7 days) before they are removed. Efficiency and success would be increased by locating the incubation and chick-rearing facilities near the wild population, i.e., on or near the McCandless Ranch. Also, what is learned from suggested modifications of the current method of incubation at Olinda should improve the hatchability of both wild and captive-produced eggs. Infertile eggs, dead embryos, and dead chicks that are collected and examined can provide important information for understanding the effects of inbreeding on reproduction and the role of artificial incubation and nutritional influences on embryo and chick development.

Recovery programs around the world have yielded information relevant to 'Alala preservation efforts. For example, wild populations of many threatened species have declined to extremely low numbers, yet have responded positively to well-conceived and carefully implemented programs. Many seemingly hopeless cases turn out to be salvageable when imaginative research and conservation programs are implemented in conjunction with concerted efforts to preserve and restore natural habitat.

Intensive in situ manipulative techniques can often lead to more effective and less-expensive augmentation of wild populations than ex situ, or captive-breeding programs, although both approaches might be required initially until the number of wild breeders is increased sufficiently. Many recovery programs have enhanced the fecundity or survivorship of wild birds considerably through the judicious application of predator control, supplemental feeding, and multiple clutching.

There are now several cases in which the geographic range of an endangered species that had been reduced to single, relict population has been expanded geographically through deliberate releases into suitable habitat. Some species, such as the Mauritius Kestrel (*Falco punctatus*), have even been successfully induced to colonize habitats that differ from their original ones. Most programs have deliberately focussed on re-establishing multiple populations to minimize the possibility of extinction as a result of local catastrophic events.

Because suitable habitat is necessary for the re-establishment of wild, self-sustaining populations, habitat preservation or restoration must have high priority in every recovery program. In the case of the 'Alala, that is especially true--essentially no native habitat remains in its pristine condition. Restoration and protection of forest preserves will have the added benefit of rescuing numerous other endangered plants and animals from similarly precarious situations.

Success in preservation of endangered species is possible only when the right combination of financial support, people, and techniques can be sustained. Although the committee recognizes that its recommendations will require the commitment of additional financial resources, it did not determine what those costs will be. But it is known that a successful recovery program will require funding continuity, effective administrative organization and commitment, creative and dedicated staffing, and effective communication and cooperation among the concerned parties. Most of those programmatic requirements are obvious, but the critical importance of coordination and cooperation needs to be emphasized. Recovery programs are never conducted in a vacuum, and they typically affect, directly or indirectly, diverse interest groups represented by government agencies, academic institutions, conservation organizations, and private land-holders. Without the support and cooperation of all those parties, decision-making will usually be determined in the political rather than the biological arena, and recovery actions may be delayed or thwarted.

EXECUTIVE SUMMARY

In summary, on the basis of the success of some efforts to restore other avian taxa, the committee believes that a successful recovery program that will allow the ‘Alala population to increase in size and distribution can be designed and implemented.

1

INTRODUCTION

The Hawaiian Crow (*Corvus hawaiiensis*), or 'Alala, is one of the most endangered birds in the world and faces likely extinction unless recovery actions begin soon. The only known wild population of this species is on private land, the 64,000-acre McCandless Ranch on the southwest slope of Mauna Loa in central Kona on the island of Hawai'i. The wild flock has been estimated recently to number 11 birds (J. Engbring, pers. comm., 1992). Little is known about the biology and life history of the few birds remaining. Because of concern about the most appropriate method to preserve the 'Alala, McCandless Ranch policy over the last decade has been to limit access to the crows by restricting entry onto the property. On the basis of trends in the overall population of the 'Alala in its historical range over the last 50 years, the U.S. Fish and Wildlife Service (USFWS) is of the opinion that the bird will become extinct unless appropriate actions are taken soon.

A captive-propagation program was established in 1976 and has been supported by the federal government and the state of Hawai'i. It was the first such program for any member of the corvid family, and it has met with some failures and some successes. The captive flock now consists of 10 birds--six that were bred and raised in captivity and four that were brought in from the wild. It is believed that the captive flock faces a severe genetic bottleneck: it produces fertile eggs each year, but very few chicks hatch. The low viability of eggs has been attributed by some biologists to inbreeding, disease, or unsatisfactory captive propagation techniques. Some captive-propagation specialists believe that genetic stock from the wild must be added to solve the problem; others disagree.

Because the 'Alala is on the endangered-species list, it is covered under the Endangered Species Act. Responsibility for implementation of recovery actions under the Endangered Species Act rests with USFWS. USFWS asked the National Research Council's Board on Biology to review available information on the 'Alala to determine the steps that would be appropriate to ensure the survival of the species. The committee on the Scientific Bases for the Preservation of the Hawaiian Crow--consisting of experts in ornithology, conservation biology, population biology, and captive propagation--was thereupon formed and charged to analyze the data and prepare a report detailing its findings, conclusions, and recommendations concerning the recovery of the 'Alala population. The committee and two special advisors examined the relevant scientific data, held a public meeting, and spoke to numerous persons, including state and federal biologists, staff at the captive-breeding facility at Olinda on the island of Maui, and owners of the McCandless Ranch. The committee also visited the Olinda facility and spent 2 days on the McCandless Ranch.

This report reflects the committee's analysis of all available information on the ‘Alala. Chapter 2 summarizes what is known and identifies what is not known about the wild population. It includes a discussion of possible direct and indirect factors that have contributed to the decline of the ‘Alala, a demographic analysis of the current wild population based on existing data, and an estimate of times to extinction of the ‘Alala and other insular populations of corvid species. The demographic analysis revealed new information about the structure of the population. Chapter 3 discusses genetic considerations regarding small populations and their role in decision-making about the recovery of the ‘Alala. Chapter 4 outlines the goals of a captive-breeding program, the history of the captive-breeding program for the ‘Alala, and an analysis of the captive population's demography and genetics. Chapter 5 describes successes achieved in the reintroduction and translocation of other endangered species and presents evidence of reasons for optimism about the recovery of the ‘Alala. Chapter 6 lays out eight options for the management of the two subpopulations of the ‘Alala, and Chapter 7 describes the committee's major findings and recommendations.

2

HISTORY OF THE WILD POPULATION AND CAUSES OF ITS DECLINE

This chapter reviews what is known about the special ecological requirements of the 'Alala, for example, its original habitat and geographic range, food requirements, and direct and indirect factors that probably influence the number of birds. Data on the demographics of other island populations of corvids, though not perfectly analogous to the situation in Hawai'i, are also considered in an attempt to estimate the probability of extinction of the 'Alala.

Each endangered species presents its own array of questions that need to be answered before any action can be prescribed and taken. Although the questions can be general, the uniqueness of each species requires that some answers be specific. In the case of the 'Alala, useful and pertinent information is sparse, and it is probably too late to obtain certain types of information. Despite the fragmentary nature of the information, a species-specific judgment must be made now concerning both the environmental factors that have contributed to the 'Alala's decline and the recovery actions that are most likely to promote the re-establishment of a self-sustaining wild population.

Past and Present Distribution of the Wild Population

The ancestral origins of the 'Alala are unknown, but it was probably derived from colonization of a *Corvus* ancestor from the Australasian region. Two recently extinct species on the islands of O'ahu (*C. impluviatus*), and O'ahu and Maui (*C. viriosus*) have been described; they are easily distinguished from the 'Alala by bill shape and body size (Olson and James, 1982b; James and Olson, 1991). *C. hawaiiensis* has a short, moderately arched bill, *C. impluviatus* had a high, impressivley arched bill, and *C. viriosus* had a long, relatively straight bill. The 'Alala has the smallest body of these species, *C. impluviatus* the largest. Whether the three species were derived from a single ancestral colonizing species or several has not yet been determined (James and Olson, 1991).

Banko and Banko (1980) compiled a detailed record of distribution and published and unpublished sightings of the 'Alala. The earliest specimen was taken during the visit of Captain Cook in 1778. Sightings and specimens obtained later by naturalists and collectors clearly reveal that the range of the 'Alala was relatively restricted on the island of Hawai'i even before its historical decline (Wilson and Evans, 1893; Perkins, 1903). Specimens collected between 1887 and 1902 were all taken on the island in a narrow V-shaped belt of dry woodlands and ‘ohi‘a forests and moist-‘ohi‘a koa forests at elevations of 500-1,800 m (Perkins, 1893; Banko and Banko, 1980). The belt extended from Hualalai on the northwestern side of the island to the southern and western sides of Kilauea Crater, perhaps to Kipuka Puaulu, which is now part of Hawai‘i Volcanoes National Park (Henshaw, 1902) (Figure 2.1). There were a few sightings in that area in the 1960s and 1970s. The species might have occupied high-elevation forests along the Hamakua coast on Mauna Kea, but disappeared from these forests before ornithologists explored there. The restricted distribution of the ‘Alala has never been fully explained and remains an enigma, but it suggests a species with rather narrow habitat requirements or preferences.

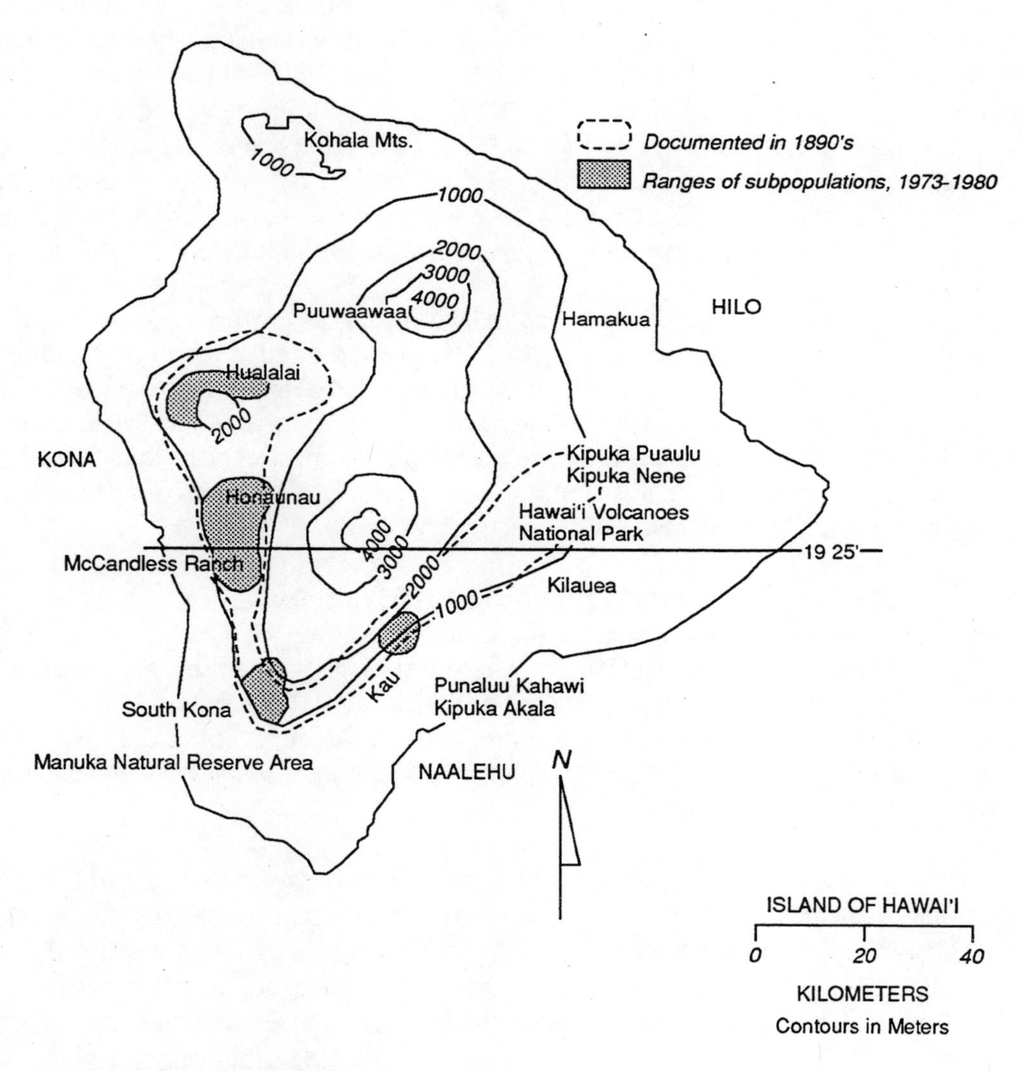

Figure 2.1 Map of the island of Hawai‘i

The Geological Setting

The distribution of the 'Alala has been almost entirely confined to the slopes of two active shield volcanoes, the north and southwestern slopes of Hualalai and the southwestern slope of Mauna Loa. Some facts about the recent geological history of these volcanoes might aid in understanding the special ecology of past and present populations of the species. Radiocarbon dating of many lava flows shows that the surface of Mauna Loa, including the entire Kona coast, is being replaced at a rate of 40% per 1,000 years (Lockwood and Lipman, 1987) and the surface of the slightly older Hualalai at 25% per 1,000 years (Moore et al., 1987). Those high rates of replacement mean that only very small areas have forests that are more than 4,000 years old (Figure 2.2). As molten lava emerges from high-altitude rifts, it flows down toward the sea as a fiery river that destroys the vegetation in its path. Two predominant types of lava are a'a (rough) and pahoehoe (smooth). Detailed maps of historical and prehistoric flows show an intricate stripe-like pattern of parallel and partially overlapping flows of varied widths. Figure 2.2 documents the ages of surface flows on Mauna Loa. On Hualalai, the situation is essentially similar, but with somewhat older flows. In the high-rainfall areas, each flow appears to have been followed fairly rapidly by revegetation; a substantial forest with relatively large trees is attained in about 200 years, depending on the type of lava being colonized.

Several recent observers have noted that 'Alala appear to prefer the interface between intact forest with tall trees and areas that are open. Before the establishment of ranches, such habitat would have been characteristic of the contact zone between forested older flows and newer, open flows. Clearly, the concept of a stable "ancient forest" habitat for the crow anywhere on this entire slope of the island is not valid. It is almost certain that former 'Alala populations continually adjusted their distribution in response to shifts in ecological conditions imposed by the periodicity of vulcanism and forest succession.

Extensive studies of the modes of natural colonization of new lava flows by plants have been made on a transect on the southeastern slope of Mauna Loa (Atkinson, 1970; Mueller-Dombois et al., 1981; Mueller-Dombois, 1987). These resulting data can serve as useful models for the understanding of the ecosystems associated with new lava flows on the Kona coast, i.e., the habitat of the 'Alala. In particular, one can estimate the characteristics of the bird's original habitat before the clearing of the forests for ranching and logging. Very few relevant areas on the Kona coast are still being colonized naturally.

When viewed in detail, each new bare lava flow is affected by unique sets of colonization events; these often result in differences in frequency of species composition from one new flow to the next. For example, the composition of major tree species is obviously similar to and derived from adjacent, older flows, but the vagaries of chance colonization, often from a distance, results in a complex mosaic of frequencies of tree species and their accompanying climbing and understory plants. Some of them are especially important as food plants for the 'Alala (see below).

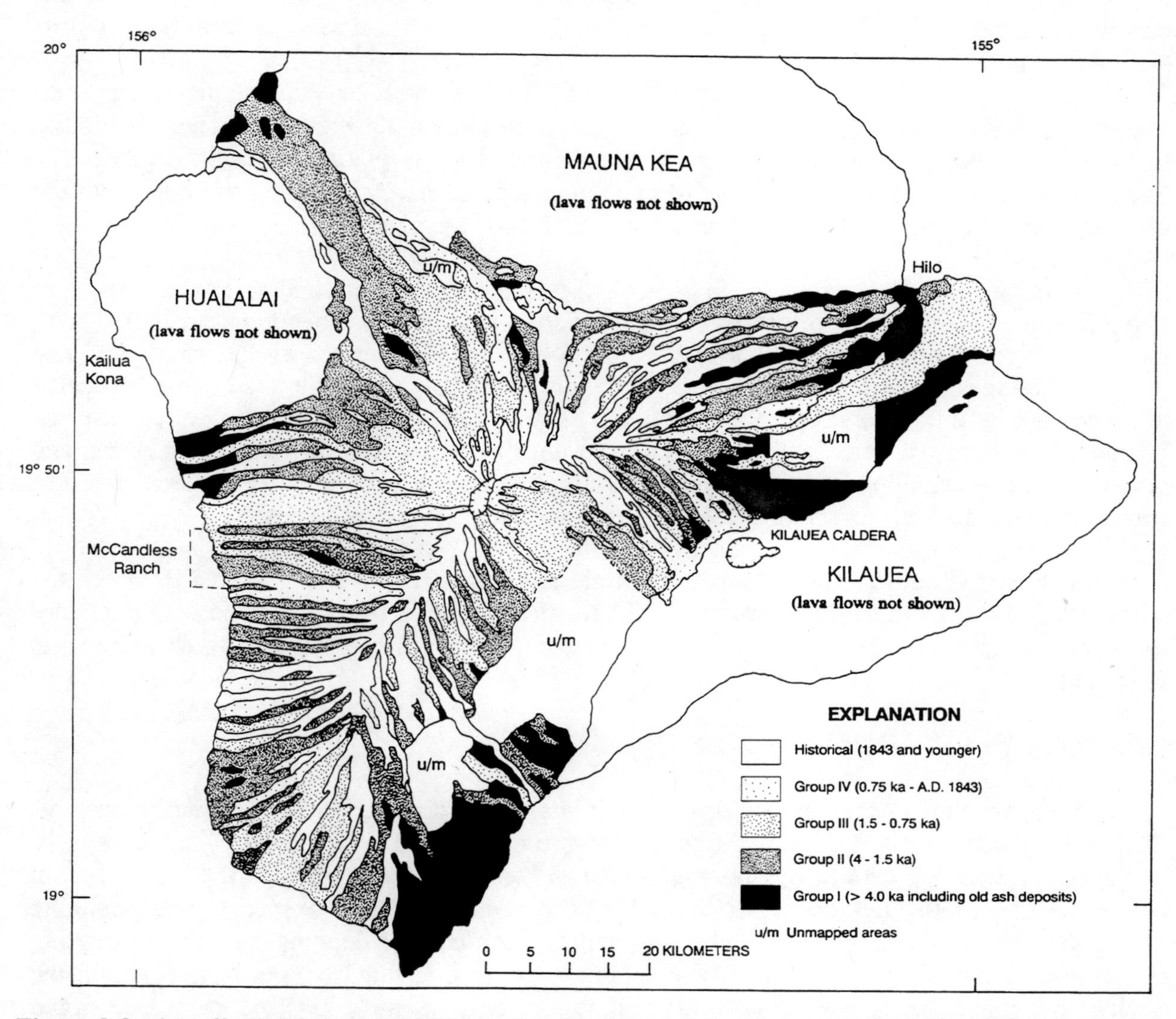

Figure 2.2 Age distribution map of Mauna Loa lava flows, showing in particular the natural parallel arrangement of the flows in present and former ʻAlala habitat on the Kona coast. Most flows are more recent than 4,000 years ago. Greatly generalized from 1:24,000 mapping; adapted from Figure 18.2 of Lockwood and Lipman (1987). ka = 1,000 years before present.

As newer raw flows repeatedly cover partially vegetated older ones, the surface pattern includes various surviving pockets of older vegetation growing on an earlier flow at a slightly lower level. Such areas are called "kipukas"; some of these can be seen on large scale maps like Figure 2.2 (note dark areas), but many small-scale events are similar and generally much more significant on the local scene. Kipukas are obvious when seen totally isolated by new raw flows, but are less conspicuous as these younger flows become newly colonized. All this greatly increases the complexity of these ecosystems' mosaics. Many native species, furthermore, display exuberant genetic variation as they colonize, sometimes leading to new populations that can be recognized as genetically distinct (Mueller-Dombois et al., 1981; Carson et al., 1990).

A prominent feature of the original flora of these evolving ecosystems is the great scarcity both of mature soil and of endemic grasses, especially at the elevations where the ʻAlala was originally found. It is clear that the ʻAlala originally had no relationship to grasslands. The bare, rough, soil-less lavas and ashfalls were colonized naturally by a number of understory and climbing plant species that are adapted to germinate on these areas.

Ranching practice has been to remove the understory plants and trees that are close together and to plant exotic forage grasses. In this long-term process, begun two centuries ago, many weed grasses became established with the forage grasses desired by the ranchers. The process also encourages many other invasive, exotic weeds. Even though it is common practice to leave some tree overstory, the resulting pastures effectively discriminate against the maintenance of native herbs and shrubs. Cattle not only browse down the native plants and grasses, which lack spines or chemicals that naturally protect them from herbivores, but also continue to spread both forage and weed grasses and larger weeds effectively.

The spreading novel grassland forms tight mats in thin soil over the lava in which the native food plants of the ʻAlala cannot germinate. Recovery and renewal of native understory can occur in an area only if cattle are excluded and a program of grass and weed removal is instituted.

Geographic Distribution of the ʻAlala

Late in the nineteenth century, the ʻAlala was reported to be "abundant locally" (Henshaw, 1902; Perkins, 1903; Munro, 1944; Banko and Banko, 1980) and "extremely numerous" (Henshaw 1902). "Hundreds" were seen in some ranch areas in the Kona District as late as the 1930s, but the observation by Perkins (1903) that the species was declining throughout its range turned out to be prophetic. Commercial logging and the continuing conversion of forest to agriculture and ranching were accompanied by steady declines in the number of crows. Many were shot as agricultural pests in the late 1800s and early 1900s, and this unnatural mortality undoubtedly contributed to the species' decline in some areas (Munro, 1944; Giffin et al., 1987). Unlike the American Crow (*Corvus brachyrhynchos*), which thrives today in farmland, the ʻAlala disappeared from highly disturbed areas--a further suggestion of relatively specialized habitat requirements.

In the Kaʻu District on the southeastern section of the island, the number of ʻAlala declined steadily throughout the 1900s (Munro, 1944; Baldwin, 1969; Banko and Banko, 1980). Baldwin (1969) did not encounter ʻAlala during a 3-day field trip between Kipuka Akala and a point northeast of Punaluʻu Kahawi, including Kipuka Nene, although some birds were still reported in this area. Three ʻAlala were encountered in open-canopy ʻohiʻa-koa forests with mixed native shrub understories during an extensive survey of the Kaʻu Forest Reserve by USFWS in 1976 (Scott et al., 1986), and a single bird was recorded in the Hawaiʻi Volcanoes National Park in 1977 (Banko and Banko, 1980). The Kaʻu District subpopulation is now believed to be extirpated, although to our knowledge no additional attempts have been made to locate ʻAlala within the district.

The history of the ʻAlala in the Kona District, on the western side of the island, has been well documented, especially since the early 1970s, and might afford insights into factors that are contributing to its critical status. Figure 2.3, compiled mainly from data supplied to the committee by the state of Hawaiʻi (J.G. Giffin, pers. comm., 1991), summarizes the changes. Between 1969 and 1976, W. E. and P. C. Banko made extensive population surveys of the Kona District for USFWS (Banko, 1974, 1976; Banko and Banko, 1980). The state of Hawaiʻi, USFWS, and others have continued to monitor the population since then (Giffin, 1983). Joint federal and state transect surveys were conducted in the Kona area from 1983 through 1987, except on the McCandless Ranch, where access was denied by the owner (Scott and Kepler, 1985; Scott et al., 1986; Giffin et al., 1987). In the Puʻuwaʻawaʻa Ranch area, on the north slope of Hualalai in north Kona, the ʻAlala were once common in dry ʻohiʻa-koa forest. In the 1940s, ʻAlala occurred in the wet forests up to the summit of Hualalai (Baldwin, 1969; Berger 1981). But from the beginning of the counts made in the 1970s, it became clear that the ʻAlala was declining precipitously in this area (Figure 2.3.a). The ʻAlala population on Hualalai declined from 26 birds in 1974 to the single banded female that was seen in March 1991 adjacent to the Hualalai Ranch; the last nest was found in 1983 (Giffin et al., 1987; J. G. Giffin, pers. comm., 1991). Census data from the Honaunau Bishop Estate Lands in central Kona parallel those from Hualalai (Figure 2.3). The only exception to the decline appears to be on the privately owned McCandless Ranch, also in central Kona, where the only known wild population of ʻAlala occurs. The owners of the ranch have reported that 10-25 birds have been present for the last decade. They permitted very little access to the property between 1980 and 1989. In 1989 and 1990, single groups of nine and four birds, respectively, were seen and videotaped on the McCandless Ranch by USFWS biologists. In January 1992, at least eight birds were seen by members of this committee on the McCandless Ranch. Surveys conducted in April 1992 have determined the current wild population on McCandless Ranch to be 11 birds (J. Engbring, pers. comm., 1992). The breeding range of the crow appears to have been further constricted since the mid-1970s (J. Engbring, pers. comm., 1992).

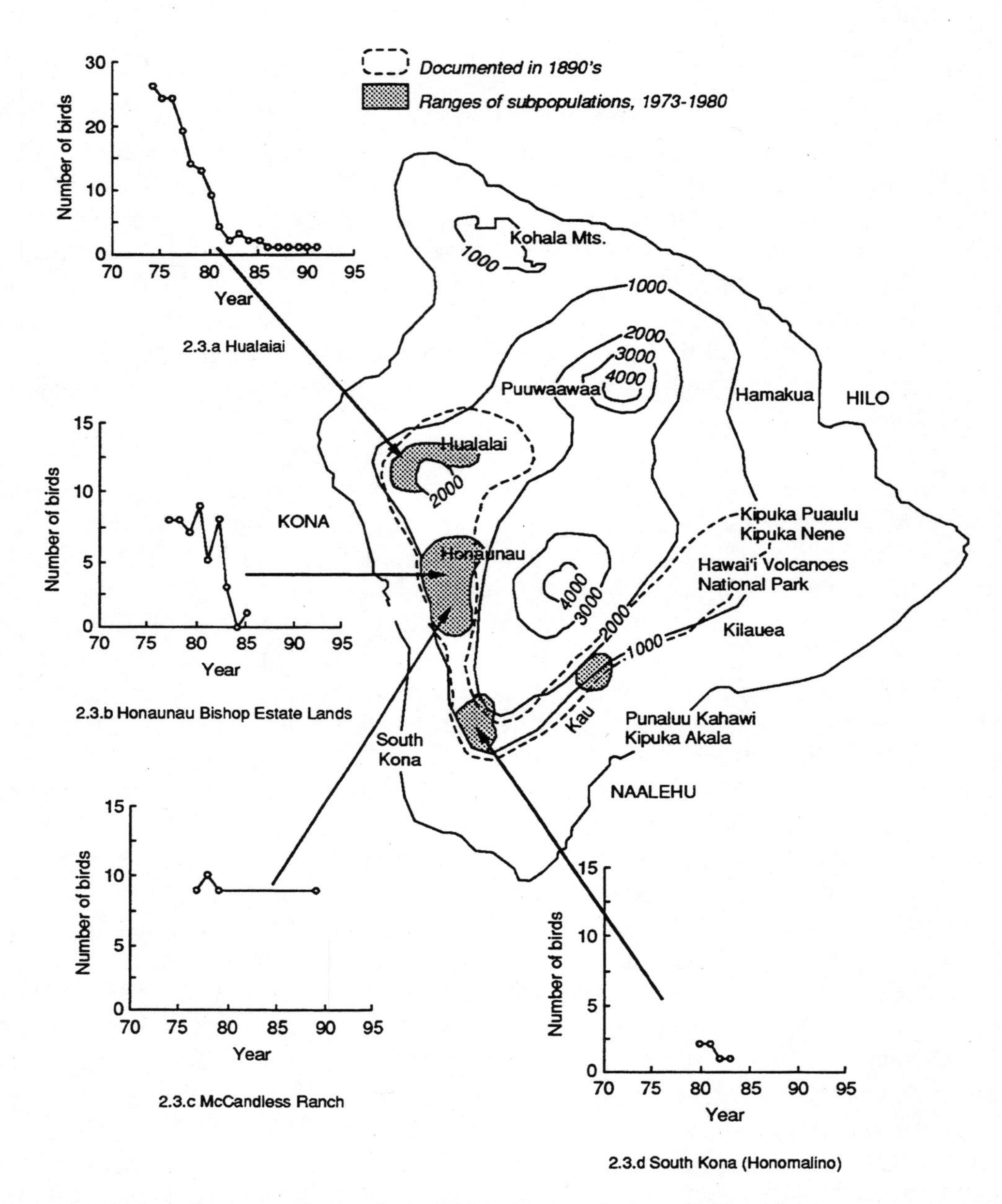

Figure 2.3 Distribution of ‘Alala from survey data collected by J. Giffin (1991).

In summary, the ‘Alala was declining but was still common in a narrow forest belt on the leeward southern and western side of the island of Hawai‘i until about 1940. The Federal ‘Alala Recovery Plan (Burr et al., 1982) gives three locations in the Kona District where at least one pair nested in the 1970s: Hualalai (north Kona), Honaunau (central Kona) and south Kona (see Figure 2.3). Those three and a location in the Ka‘u Forest Reserve, also known to be in the ʻAlala's historical range, were designated as essential habitat to be protected and managed for the ʻAlala. Since the Alala Recovery Plan was issued, the population has continued to decline. With the possible exception of a single female in north Kona, the small remaining wild population is now believed to be restricted to a single location in central Kona--the McCandless Ranch.

Assessment of Extrinsic Causes of the Decline in the Population

A decline in the numbers of any population is a function of low rates of reproduction, low rates of survival, or both. This section explores factors that are likely to be affecting those rates and discusses the particular ecological requirements of the species. As with many other endangered species, many of the factors associated with the decline of the ʻAlala have been proposed as causes. Each of these factors could still be contributing to the decline, but it is also possible that the factors operating now are different in kind and magnitude from those which were most important in previous decades. Because many factors are likely to have acted in concert, it is not possible, on the basis of the scientific data now available, to determine the extent of their individual contributions.

Figure 2.4 is a graphic summary of environmental factors that are known to affect the numbers of ʻAlala.[1] The envirogram shows that activities of humans are likely to have negatively affected the survival and reproduction of this species in various ways. If data were available on the rates of the effects of each of the 10 direct and 9 indirect factors and all the important factors had been identified, we could say what is causing the decline of the ʻAlala; however, that information is not available. The following discussion details what is known about the factors.

Habitat and Food

The first two directly acting factors are habitat and food: the ʻAlala inhabits fairly closed (canopy cover more than 60%) to moderately open moist ʻohiʻa-koa and wet or dry ʻohiʻa forests with a diverse understory of fruit-bearing trees and shrubs. Recent records of ʻAlala note nest sites in wet forests (Giffin, 1983; Giffin et al., 1987). In wet forests, tall, emergent koa trees serve as lookout posts and resting and displaying sites for the birds. All natural-history data and banding results indicate that ʻAlala are permanently territorial and extremely site-faithful once they become breeding adults. Treetop sitting is undoubtedly related to territorial advertisement and defense, as in other corvids (Kilham, 1985a,b; McGowan and Woolfenden, 1989). Nests are usually built in the upper branches of ʻohiʻas. In the current forest belt of the Kona District,

[1] Figure 2.4 is a modification of the stylized "envirograms" recommended by Andrewartha and Birch (1986) for summarizing information about direct and indirect factors affecting a particular population.

koa trees are restricted to a narrower and higher range than are ‘ohi‘as, usually above 1,500 m. Crows move short distances up the slopes of Hualalai and Mauna Loa into this region during the breeding season (Giffin, 1983). Seasonal movements to the lower elevations in the nonbreeding season appear to track the fruiting periods of the principal food plants (Giffin et al., 1987).

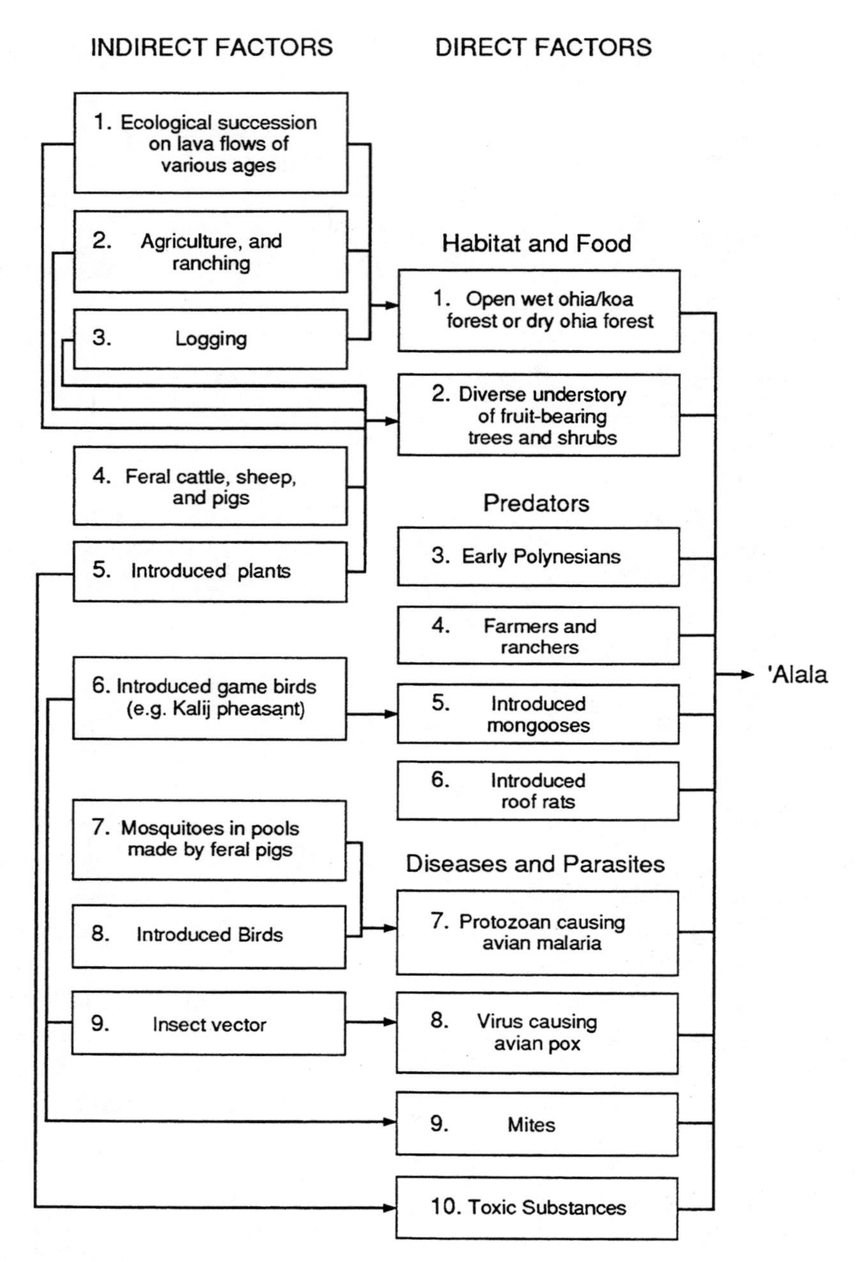

Figure 2.4 Factors influencing numbers of ‘Alala

HISTORY OF THE WILD POPULATION AND CAUSES OF ITS DECLINE

At least five indirect factors affect the suitability of both habitat and food sources for the ʻAlala. The first is the impact of lava flows in the ʻAlala's historical range, resulting in alterations to which the bird must continually adjust.

Other indirectly acting factors are habitat changes caused by agriculture, ranching, and logging. A prevailing hypothesis is that clearing of land by logging and, more important, by cattle and sheep grazing has so altered the available food supply that fecundity and survivorship have been severely affected (Banko and Banko, 1980; Giffin et al., 1987). It cannot be disputed that outright conversion of ʻohiʻa-koa forest lands into open pasture has greatly reduced the total amount of forest habitat throughout the original range of ʻAlala (Burr, 1984; Scott et al., 1986; Giffin et al., 1987). Widespread clear-cutting and grazing also fragmented the formerly continuous ʻAlala population into numerous smaller subpopulations, each of which then became susceptible to additional perils (see Chapter 3 for a discussion of small populations). Thus, clearing of forests to produce grazing lands has clearly caused substantial reductions in the numbers of wild ʻAlala (Banko and Banko, 1980; Giffin et al., 1987).

In the first half of the 1900s, substantial portions of the lower-elevation forests along the Kona and Kaʻu coasts were developed into pastures, orchards, and settlements; by the 1940s, the ʻAlala was essentially extirpated from these areas (Banko and Banko, 1980). Similarly, the disappearance of crows from most of their former range in central Kona was associated with heavy logging of valuable koa trees on the slopes of Mauna Loa from the mid-1920s to the late 1950s. Dry forests at low and middle elevations were significantly modified by the harvesting of sandalwood (*Santalum* spp.) in the 1800s, as well as by the later spread of exotic fountain grass (*Pennisetum setaceum*) and periodic wild fires. The last ʻAlala disappeared from the dry forests on the north slope of Hualalai in 1964 (Tomich, 1971). Logging was often accompanied by the introduction of pigs, goats, sheep, and cattle, and grazing further modified the forest by preventing regeneration of koa trees and eliminating many native understory plants (Ralph and van Riper, 1985).

Much of the food of the ʻAlala consists of fruits and invertebrates taken from the forest understory (Banko, 1976; Scott et al., 1986). Many of the understory plants and climbing vines are known to be important sources of fruit eaten by the ʻAlala. Although ʻAlala are omnivorous, fruits account for 33-66% of the adult diet (Sakai and Carpenter, 1990). Fruits of historical importance include ʻieʻie (*Freycinetia arborea*), ʻolapa (*Cheirodendron trigynum*), ʻoha-kepau (*Clermontia* spp.), oha (*Cyanea* spp.), mamaki (*Pipturus albidus*), pilo (*Coprosma* spp.), akala (*Rubus hawaiiensis*), and lama (*Diospyros* spp.). Seeds of manono (*Gouldia terminalis*) and hoʻawa (*Pittosporum hosmeri*) are also used (Burr et al., 1982). The introduced banana poka (*Passiflora mollissima*) has become a major item in the diet. In addition, isopods, arachnids, and land snails are taken from understory trees and shrubs or gleaned from branches and clumps of vegetation in the upper canopy. The diet fed by ʻAlala to their nestlings includes these items plus nestling songbirds and mice (Perkins, 1903; Munro, 1944; Sakai and Ralph, 1980; Sakai et al., 1986; Scott et al., 1986).

Another indirect factor that affects the habitat and food of the ʻAlala is the common practice in Hawaiʻi of introducing feral ungulates into the forest. Both on private land and on land leased by the state, cattle, pigs, and sheep have occupied the forest belt for many years, often with considerable impact (Stone and Loope, 1987). Where the animals are overstocked, their foraging destroys native understory plants and deprives the ʻAlala of many of its sources of fruit.

Since 1950, the state of Hawaiʻi has had a habitat-improvement program. Some areas that had been severely degraded by timber harvest and feral ungulates are recovering owing to the removal of ungulates, and that is the case in the Hualalai Reserve established in 1984. Cattle had been present in the wet forest of Hualalai for the previous 100 years. On the drier northern slopes in the Puʻuwaʻawaʻa Reserve, from which cattle and pigs have also been removed, the koa trees have begun to recover. Bird surveys in this area and in the Manuka Natural Area in south Kona, however, have continued to show declines in the numbers of ʻAlala and other native birds (Giffin, 1983; Sakai et al., 1986; Giffin, 1990, 1991).

Ranching practices and the behavior of foraging cattle commonly result in reduction and even elimination of understory and subcanopy vegetation, and ground vegetation is replaced with introduced forage grasses and weeds. Even on ranches where mature trees are left in abundance, the resulting parklike environment often supports a greatly reduced diversity of plant species, in which natural regeneration of canopy trees is dramatically reduced (Baldwin and Fagerlund, 1943). Understory vegetation of the native forest included a rich diversity of fruit-producing species (e.g., Lobeliaceae and Rubiaceae). Many of those plants are eaten preferentially by grazing cattle and are uprooted by feral swine; when fully exposed and accessible to livestock and pigs, they become scarce and patchy. Even on the McCandless Ranch, where ʻAlala are still found, fruiting species of the understory are widely scattered as individual plants, frequently growing out of fallen logs or tree crotches above easy reach of cattle (observations by this committee, January 1992). Under such conditions, these understory species have greatly reduced capacities to regenerate, recruit into dense stands, or recolonize areas from which they have been removed.

Biologists familiar with the plant life of the Kona slopes unanimously report a strong negative association between cattle and several of the most important food plants of the ʻAlala. That is especially true for the ʻieʻie, which was shown in early twentieth century photographs to have covered huge areas of ʻohiʻa forest in dense tangles. They were visited by ʻAlala in great numbers, and crows were said to have nested directly among the vine tangles (Perkins, 1903). Evidence exists that several Hawaiian bird species, including ʻAlala, were important pollinators of ʻieʻie (Cox, 1983); substantial loss of such important food sources could seriously alter time and energy budgets, and adversly affect reproduction and survival of the ʻAlala.

As a direct result of logging and ranching, essentially no pristine example of the closed to moderately open moist ʻohiʻa-koa and wet or dry ʻohiʻa forest zone exists anywhere on the

Kona slopes. As of 1992, most of the remaining stands of intact humid forests are at the lower limits of ʻAlala's modern range and consist of moist ʻohiʻa forest growing on aa lava flows. These stands were used by ʻAlala for late summer and winter foraging, but not as primary breeding habitat, which appears to be at an elevation of 1,000-1,600 m.

A dramatic correlation between loss of understory habitat and decline of an ʻAlala population occurred in the late 1970s on Hualalai (Giffin, 1983; Carson, pers. comm. 1992). According to Giffin (1983; p.24), "one of the most important ʻAlala nesting areas on Hualalai was partially cleared with bulldozers in the fall of 1976. Understory vegetation was removed and exotic pasture grasses were planted. Continuous grazing by livestock during the next 4 years further opened the forest floor. The number of active crow nests in that area decreased from four in 1977 to none in 1980."

The persistence of ʻAlala on one particular ranch does not in itself mean that this property provides the only genuinely optimal habitat remaining within the historical range--only that the habitat there remains suitable and might be less degraded than other areas of former habitation. The McCandless Ranch lies directly at the center of the region where ʻAlala historically was most common (middle elevations of central Kona). Because small populations are subject to extinction from chance events and because ʻAlala are sedentary and unlikely to colonize new areas quickly, there could well be other suitable habitat areas within the historical distribution that are simply unoccupied.

Another complicating aspect is that ʻAlala began to decline before 1900 in areas not dramatically altered by ranching, and throughout this century they have persisted in habitats where cattle ranching has been the dominant land-use activity. Numerous anecdotal reports and observations indicate that ʻAlala in recent decades have preferred forest habitats that are broken by clearings and edges of habitats, of the kind typical on forested cattle ranches throughout middle and upper elevations of the Kona region (e.g., Berger, 1981). However, the overall record shows that past land-use practices have severely degraded the forest and in many areas removed aspects of the native forest that the bird might require, and the degradation is likely to have been a major factor in the overall decline of the ʻAlala (Tomich, 1969, 1971; Banko, 1974; Burr et al., 1982; Giffin 1983; Sakai et al., 1986; Giffin et al., 1987). The fact that apparently suitable native forest is now unoccupied by ʻAlala is difficult to explain (Banko, 1974) other than by suggesting that it occurs in fragments that might not be discovered by such a basically sedentary, traditional species or that the fragments are too small or too low in elevation to provide sustenance throughout the year.

Predators

Direct factors 3-6 in Figure 2.4 are predators that directly affect the numbers of ʻAlala: early Polynesians, farmers and ranchers, introduced mongooses (*Herpestes auropunctatus*), and introduced roof (Polynesian) rats.

Polynesians used crow feathers in some of their ceremonies, but not to the extent that they used feathers of the brightly colored Honeycreepers (Drepanidinae spp.). European settlers are known to have hunted crows regularly for sport. When crows were common, they entered poultry yards, and many were shot by farmers as agricultural pests (Banko and Banko, 1980; Munro in Berger, 1981; Burr et al., 1982). Continuing, but unsubstantiated, reports of 'Alala shootings in recent years in Hualalai and other areas of north Kona despite legal protection (J.G. Giffin, pers. comm., 1991) point to the need for increased enforcement and public education. Considering the small size and relict distribution of the extant population, even a low mortality stemming from shooting could seriously hamper recovery efforts.

The Hawaiian Islands have no native mammals except bats, and there is no evidence that native hawks or owls have constituted a problem as predators of the 'Alala. But two introduced predatory mammals are abundant in the forest belt, and they take a heavy toll on the native fauna. Mongooses, introduced in 1883, are effective diurnal ground predators on young crows. Like other corvids, 'Alala fledglings leave the nest before they are strong fliers and spend some time on the ground or climbing the understory vegetation while still being fed by their parents. It is a period of high vulnerability to mongooses and presumably also feral cats (Tomich, 1969, 1971; Banko, 1974, 1976; Giffin, 1983). The McCandless Ranch, with some financial support from USFWS, has had a trapping program to control its mongoose population for the last few years (see Figure 2.5). An indirect factor that might contribute to predation on young 'Alala by mongooses is the presence in the forest of large numbers of the introduced Kalij Pheasants (*Lophura leucomelana*) and Turkeys (*Meleagris gallipavo*). Those birds are common on the McCandless Ranch (Engbring, 1990; committee observations, January 1992) and elsewhere within the 'Alala's former distribution (Lewin and Lewin, 1984); they and their eggs and young are important prey of the mongoose. Their abundance might help to support a large mongoose population, which in turn could increase the juvenile mortality of 'Alala.

The introduced roof rat is also a common predator in the forest belt (Tomich, 1969). This nocturnal arboreal predator can take eggs and nestlings. Partial brood loss from nests should probably be attributed to predation by rats (Tomich 1969, 1971). No evidence supports or refutes the idea that introduced predators are the major cause of the decline in the wild 'Alala population.

Diseases and Parasites

Factors 7-10 in Figure 2.4 are diseases, an ectoparasite, and potentially toxic substances. Two avian diseases--pox and malaria--introduced to Hawai'i after colonization by Europeans appear to have played an important role in the decline of many endemic birds (Warner, 1968; van Riper et al., 1986). Avian pox is caused by an arbovirus and is transmitted by direct contact with an infected bird, by secondary contact with a contaminated object, or mechanically by vectors, such as mosquitoes and biting flies (van Riper and van Riper, 1980; Cavill, 1982). Birds with pox usually develop lesions at the immediate site of infection in the mouth and upper

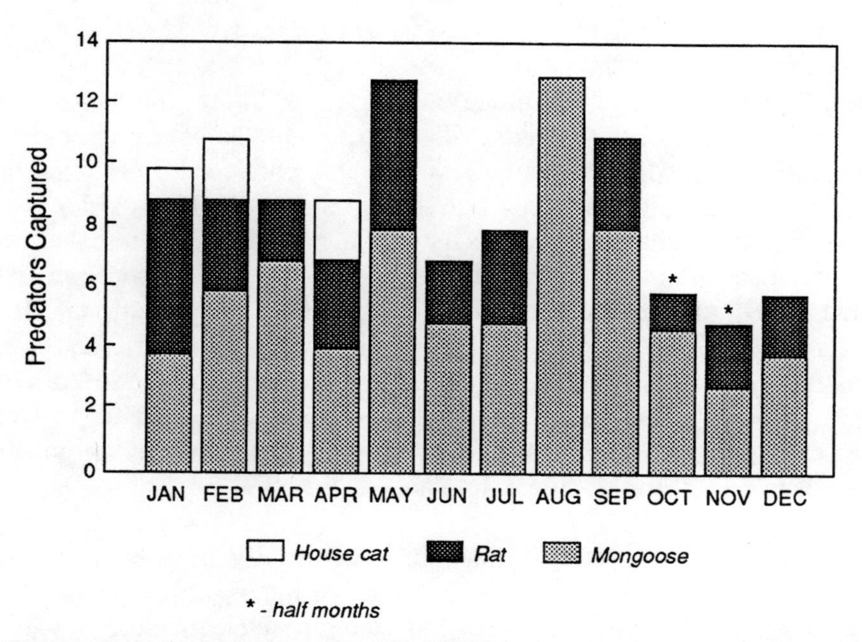

Figure 2.5 1991 Predator Trapping Results, McCandless Ranch. Information from P. Simmons (pers. comm. 1992).

respiratory tract or on the integument on the bill or legs or around the eye. Exposure to the disease is generally believed to confer immunity, but young birds seem particularly prone to severe, debilitating lesions. For example, in Galapagos Mockingbirds (*Nesomimus p. parvulus*), Vargas (1987) found a higher incidence of active lesions in nestlings and juveniles than in adults and higher post fledgling mortality in infected than noninfected birds. Nestling and fledgling 'Alala have been observed with pox-like lesions on a number of occasions (Banko and Banko, 1980; Giffin, 1983), and pox has been implicated in contributing to mortality in young 'Alala (Jenkins et al., 1989). Avian pox apparently was widespread in the native Hawaiian avifauna, including 'Alala, by the turn of the century (Perkins, 1903), and it persists today (Ralph and van Riper, 1986; Scott et al., 1986). Although the incidence rate in wild 'Alala is unknown, Jenkins et al. (1989) detected precipitating antibodies against fowlpox virus in blood sera collected from three adults (100% incidence), and two of five (40% incidence) nestlings. All three of the adult samples also reacted to the Hawaiian poxvirus isolate.

The blood protozoan that causes avian malaria (*Plasmodium relictum capistranoae*) is believed to have arrived in Hawai'i in the 1920s with introduced passerine birds (van Riper et al., 1985; Stone and Loope, 1987), and it has been identified in a variety of endemic Hawaiian birds, including the 'Alala (cf. Banko and Banko, 1980; Jenkins et al., 1989). As with avian pox, there is relatively little information on the rate of occurrence of avian malaria in 'Alala.

Jenkins et al. (1989) detected a heavy avian malaria infection in only one of 10 nestlings and in none of six adults. However, these authors noted that young birds are more susceptible to infection than adults, and that the debilitating effects of malaria would be expected to increase both direct and indirect mortality in nestlings and fledglings. The principal vector for this parasite is the introduced night-biting, ornithophilous mosquito, *Culex quinquefasciatus*, although another introduced species of mosquito, *Aedes albopictus*, can also act as a vector (Scott et al., 1986). The vulnerability of nestlings to malaria is probably enhanced by the fact that mosquito populations generally peak during the 'Alala's breeding period (Jenkins et al., 1989). Results of inoculation experiments conducted by van Riper et al. (1985) indicate that native Hawaiian birds are much more susceptible than introduced species to mortality from malaria, and Scott et al. (1986) have noted the negative relationship between the elevational distribution of mosquitoes and that of the surviving populations of native birds (see also Warner, 1968). Moreover, rooting by feral pigs, and stock ponds and troughs for cattle, provide sites for mosquito breeding, and hence the spread of malaria. The results lead us to conclude, with Ralph and van Riper (1985) and others, that avian diseases--particularly avian pox and malaria--have probably been a major factor in the decline and distribution of the 'Alala.

Banko (1974) found an 'Alala fledgling that was heavily infested with the introduced northern fowl mite (*Ornithonyssus sylviarum*). Although the significance of this parasite is unknown, it has been reported to cause mortality in young poultry (Banko, 1984). Heavy mite infestations could result from weakness or disease brought about by other factors. The extent to which introduced birds, such as the Kalij Pheasant, serve as reservoirs for disease and ectoparasites is poorly known, but certainly merits further investigation. The possibility that the seeds or fruits of introduced species of plants, such as the banana poka, that are now regularly exploited by the crow might contain potentially harmful substances also warrants further study.

Estimates of Population Viability and Time to Extinction

Approaches

The committee was charged with estimating the "minimum viable population for survival" of the 'Alala. That cannot be done, in part because no set population size is sufficient to ensure survival within the changing, fragmented landscape of the Hawaiian forests. We can only estimate probabilities of extinction as a function of population size. The number of birds is too low for long-term survival to be possible without *active* management.

What follows are an evaluation of the species' current demography in the wild and an estimate of its time to extinction if there is only *passive* management.

HISTORY OF THE WILD POPULATION AND CAUSES OF ITS DECLINE

Demographic Analyses

Given what we know about the demographics and approximate size of the wild population of the 'Alala, we built a deterministic model of the population dynamics of the 'Alala. Such a model, by definition, lacks the chance events that we know can doom small populations. Nonetheless, the exercise provides clues about the decline of the 'Alala in most areas and about its persistence on the McCandless Ranch.

Limited banding data at the McCandless Ranch (where the numbers appear not to have changed much in a decade) demonstrate that adults are long-lived, with annual survivorship approaching 90% (Banko and Banko, 1980). The species has a relatively low natural reproductive rate, with protracted periods of nestling development and fledgling dependence. As previously noted by Banko and Banko (1980), the 'Alala appears to have a smaller average clutch size than other members of *Corvus*, although current comparative information is primarily restricted to mainland species (Table 2.1). The incubation period of the 'Alala is similar to those of other members of *Corvus*, but the fledging time seems to be slightly longer than expected (Table 2.2). Pairs appear to occupy permanent ("traditional") territories of comparatively small size (about 1 km^2), which also contain nonbreeding offspring from the breeding seasons of the preceding 1 or 2 years. The social roles of the nonbreeding offspring are unknown; however, evidence suggests that they are not involved in cooperative rearing of young, which occurs in a number of corvid species (Woolfenden, 1973; Kilham, 1984a,b,c; Woolfenden and Fitzpatrick, 1984). Nonbreeders might help a breeding pair to defend their territory against conspecifics. There is, however, little evidence of general reproductive failure.

Several workers (Banko, 1976; Burr et al., 1982; Giffin, 1983) have concluded that the rate of recruitment of young birds into the adult population is unusually low in the 'Alala. Although recruitment might have been higher under the pristine conditions in which the species evolved (which included no mammalian predators) than it has been in recent decades, the data in Table 2.3 indicate that the 'Alala's reproductive success is somewhat lower than that of temperate corvids. Comparative data for other tropical corvids, however, do not exist.

Existing information on age structure and age distribution of the 'Alala is in Box Table 1. It includes data from Banko and Banko (1980), who reported 72 known nesting attempts; data on 27 banded birds spanning a period from 1973 until 1980, when banding ceased, plus resightings of these birds; and the census data in Figure 2.3. The observations, although limited in scope, allow us to estimate demographic measures, given two precautions. The first is that, until April 1992, no sustematic observations were made for nearly a decade (during the 1980s), because the remaining birds were on the McCandless Ranch, where access was not permitted. The second is that the samples of banded birds are very small. Nonetheless, many of our conclusions are consistent with observations of 'Alala behavior and population biology during the 1970s and 1980s, both on and off the McCandless Ranch.

Table 2.1 Average clutch size and range of clutch sizes reported in various *Corvus* species

Species	No. Clutches	Mean	Range	Location	Source
Corvus hawaiiensis	4	3.5	2-5	Hawai'i	Banko and Banko, 1980
	11	2.2	1-4	Hawai'i	Temple and Jenkins, 1981
C. caurinus	189	4.0	1-4	Canada	Richardson et al., 1985
C. cryptoleucus			3-8	N. America	Bent, 1946
C. brachrhynchus		4.4	2-6	N. America	Emlen, 1942
C. frugilegus	182	4.2	1-7	England	Holyoak, 1967
	151	4.3	2-7	England	Lockie,1955
C. monedula	80	4.2	1-6	Switzerland	Zimmermann, 1951
	29	5.1	3-6	Finland	Antikanen, 1978
	233	4.4	1-6	England	Lockie, 1955
	431	4.3	2-7	England	Holyoak, 1967
C. corone	28	4.1	--	Scotland	Yom-Tov, 1974
	49	4.1	2-6	Scotland	Picozzi, 1975
	168	3.9	1-7	England	Holyoak, 1967
	74	4.5	--	W. Germany	Wittenberg, 1968
	39	4.6	3-6	Finland	Antikanen, 1978
	198	4.7	1-7	Finland	Tenovuo, 1963
	39	4.6	3-6	Norway	Slagsvold et al., 1984
C. capensis	7	3.9	3-4	S. Africa	Skead, 1952
C. orru	52	4.8	1-6	Australia	Rowley, 1973
C. bennetti	21	4.2	1-6	Australia	Rowley, 1973
C. coronoides	137	4.4	1-6	Australia	Rowley, 1973
C. mellori	570	4.2	1-6	Australia	Rowley, 1973
C. tasmanicus	11	4.2	3-6	Australia	Rowley, 1973
C. corax	45	6.0	3-7	N. America	Stiehl, 1985
	67	5.2	3-7	England	Holyoak, 1967
	245	4.5	--	Wales	Newton et al., 1983

Table 2.2 Incubation lengths and fledging ages reported in various *Corvus* species

Species	Incubation Length[a] days	Fledging Age[b] days	Source
Corvus hawaiiensis	19-20	39-43	Banko and Banko, 1980
C. kubaryi	~ 21	~ 33	R.E. Beck (pers. comm.)
C. caurinus	18-19	30-35	Butler et al., 1984
C. brachyrhynchos	16-21	26-35	Emlen, 1942; Bent, 1946; Chamberlain-Auger et al., 1990
C. monedula	17-18	~ 18	Zimmermann, 1951
C. frugilegus	16-18	32-39	Lockie, 1955; Richards, 1973; Goodwin, 1976
C. capensis	18-19	~ 38	Skead, 1952; Goodwin, 1976
C. corone	17-21	30-36	Holyoak, 1967; Wittenberg, 1968; Coombs, 1978
C. macrorhynchus	17-19		Goodwin, 1976
C. orru	19-20	38-48	Rowley, 1973
C. bennetti	~ 17	29-31	Rowley, 1973
C. coronoides	19-21	40-45	Rowley, 1973
C. mellori	19-20	34-41	Rowley, 1973
C. albus	18-19	~ 43	Lamm, 1958; Goodwin, 1976
C. ruficollis	20-22	37-45	Goodwin, 1976
C. corax	18-21	38-44	Gwinner, 1965; Stiehl, 1985

[a] Although some of reported variation in incubation length probably reflects attentiveness of individual females, large proportion of variation is probably due to method. Accurate determination of incubation in the field is complicated by female corvids' period before oviposition and beginning of incubation usually before clutch is complete, so that eggs hatch asynchronously (cf. Lockie, 1955; Wittenberg, 1968; Butler et al., 1984). Because most workers have used standard method of determining incubation length, time between laying and hatching of last egg (Nice, 1954), some degree of inaccuracy is probable (see Greenlaw and Miller, 1983).

[b] Many corvids leave nest several days before fully capable of flight. Additionally, observer-caused disturbance is known to cause premature fledging (cf. Butler et al., 1984).

Table 2.3 A comparison of reproductive success reported in various *Corvus* and other species

Species	% Eggs Hatched	% Nestlings Fledged	% Nests Successful	Fledglings per Nest	Source
Cyanocitta stelleri	92	27	31	-	Ligon, 1971
C. coerulescens	54	56	53	1.5	Woolfenden, 1978
Psilorhinus morio	-	76	80	3.0	Lawton and Guinton, 1981
Pica pica	80	-	75	-	Evendon, 1947
	-	39	-	1.6[a]	Husby, 1986
	-	-	41-78[a]	1.1-1.8[a]	Dhindsa and Boag, 1990
	-	-	61	2.5	Buitron, 1988
Corvus caurinus	74	44	78	1.3	Butler et al., 1984
C. brachyrhynchus	-	33	-	-	Emlen, 1942
	-	-	-	2.0 (0.8-3.3)	Chamberlain-Auger et al., 1990
C. monedula	-	73	-	2.9b	Lockie, 1955
C. frugilegus	85	72	-	2.0-3.7[b]	Owen, 1959
	-	71	-	2.7[b]	Lockie, 1955
C. corone	46	58	43	1.2[a]	Yom-Tov, 1974
	90	19	-	-	Tenovuo, 1963
C. coronoides	72	48	54-86	1.53 (0.9-1.8)	Rowley, 1973
C. mellori	75	47	70	1.99 (0.2-2.0)	Rowley, 1973
C. orru	70	31	60	1.0[c]	Rowley, 1973
C. bennetti	72	5	14	0.1[c]	Rowley, 1973
C. corax	-	70	-	2.3 (2.2-2.5)	Stiehl, 1985
C. hawaiiensis	-	-	66	0.8 (0.5-1.3)[d]	Banko and Banko, 1980
	-	-	43	0.7 (0.5-0.8)[e]	Temple and Jenkins, 1981

[a] Data from control nests only.

[b] Unsuccessful nests excluded from calculation.

[c] Data collected during year of severe drought.

[d] Calculations based on 50 nests located and observed during 1973-1978.

[e] Calculations based on 16 nests located and observed during 1979-1980

HISTORY OF THE WILD POPULATION AND CAUSES OF ITS DECLINE

On the basis of data obtained through 1980 (Banko and Banko, 1980), the age at first breeding appears to be 2 years. Of four fledglings later found nesting, three were 2 years old, and the other was 3 years old. It is not known whether the 3-year-old breeder (Hualalai bird 012) was missed as a breeder at age 2. Young 'Alala accompany adults for up to a year (maybe more) after fledging, and delayed breeding is typical of such group-living corvids. Therefore, it is extremely unlikely that 1-year-old 'Alala breed.

By assuming 2 years to be the age at first breeding, we can represent the demography of the 'Alala as a three-stage model with several parameters (see Box Figure 1). In addition to the number of young produced per breeding adult, three survivorship parameters are estimated: adult survivorship (i.e., the proportion of the population of breeding adults surviving from year t to year t + 1, for $t \geq 2$), juvenile survivorship (proportion of recently fledged birds surviving to the end of their first year), and yearling survivorship (from 1 year old to 2 years old).

Average adult survivorship can be estimated in two ways. In the observation weighted method, birds are assumed to have died immediately after they were last seen. The number of year-to-following-year survivals is averaged across all the birds. Thus in Box Table 1, adult 025 survived 12 of a possible 12 year-to-following-year intervals. In contrast, bird 006 was not seen after reaching 2 years of age: as an adult, it survived zero year-to-following-year intervals of a possible one. Pooling these data for all birds establishes the minimum estimate of average survivorship; bird 006, for instance, might have survived several years after last being seen, but might not have been seen clearly enough to identify its color bands.

The second method for extimating survival is a time-weighted method. Of five birds known to have been adults on the McCandless Ranch, two were alive after 9 years. If yearly survival is s, then 2-year survival is s^2, 3-year survival s^3, and so on. Hence, $s^9 = 2/5$, and hence $s = 0.90$.

For estimating survivorship and reproduction of 'Alala, we separated the small pool of data on birds on the McCandless Ranch (where the population was approximately stable) from the pool of data on birds at Hualalai and Honaunau (where 'Alala were precipitously declining). *The results are different from that of most previous authors regarding the proximate causes of 'Alala decline from 1974 through 1982: reproduction and recruitment were nearly normal among breeding pairs, but adult survival and nesting attempts were exceedingly low, except on the McCandless Ranch.*

Fledgling production *per nesting pair* averaged 0.92 on McCandless Ranch and 0.87 elsewhere (Box Table 4). Juvenile survival was 50% on the McCandless Ranch and 43% elsewhere (Hualalai, Honoaunau and south Kona), and yearling survival was 90%$^+$ everywhere (Box Table 2). For modeling purposes, we use 50% and 90% as juvenile (first-year) and yearling (second-year) survivorship, respectively. None of these estimates of reproduction and subadult survival is atypical of a healthy corvid population, especially given the small average clutch size and extended nestling period of 'Alala. Yearling survival (between 1 and 2 years) is often the highest of any year, because the birds have avoided the potentially fatal mistakes of inexperience, yet do not have the responsibilities of parenthood. Those numbers also match observations, both on and off the McCandless Ranch, that show that the birds are breeding successfully and that dependent young often accompany adults throughout the year. However, an extremely important observation is that fledgling production drops to 0.66 away from the McCandless Ranch when all pairs that failed to nest (or whose failed nests went unobserved) are added to the sample (Box Tables 3 and 4). Even off the McCandless Ranch, however, there was no "general reproductive failure" of the kind postulated by many previous authors (Banko and Banko, 1980; Burr et al., 1982; Burr, 1984; Jenkins et al., 1989). Some 'Alala were breeding, often successfully, even during the period when the population was crashing.

The sharpest contrast between the McCandless Ranch and elsewhere appears to be in *adult survival* (Box Table 2). On the McCandless Ranch, annual survival was between 80% and 90%, which is typical of corvids. At Hualalai and Honaunau, the minimum estimate of adult survival is 66%. A time-weighted estimate is more difficult, because birds were banded over a 6-year period. However, only one of 10 'Alala (Hualalai bird 025) is known to have survived more than 2 years as an adult! Seven 'Alala were alive as young adults in 1980. Had they survived, all seven could have been recorded during a thorough census of these regions in 1986. Only bird 025 was alive after these 6 years; the time-weighted estimate of annual survival is about 72%.

Except on the McCandless Ranch, an average of one-third of all remaining adult 'Alala died or disappeared annually between 1975 and 1986. This conclusion is entirely consistent with the observed rate of decline in 'Alala numbers from census counts at Hualalai and Honaunau during that period (Figure 2.3). This adult survivorship is extremely low for a corvid.

Florida Scrub Jays (*Aphelocoma c. coerulescens*) share many of the social and demographic characteristics of 'Alala. They are sedentary, omnivorous corvids living in permanently defended territories, and they exhibit delayed reproduction (Woolfenden and Fitzpatrick, 1984, 1990). Offspring remain with the parents for one or more years, creating a "standing crop" of nonbreeding individuals that constantly interacts with the population of breeders and survive at rates different from those of the breeders (Fitzpatrick and Woolfenden, 1986). Annual survivorship of breeding jays averages about 80% and has dropped below 70% only once in 23 years, during a catastrophic epidemic (Woolfenden and Fitzpatrick, 1991). Adult survival between 80% and 90% is reported among Santa Cruz Island Scrub Jays (*Aphelocoma coerulescens insularis*) (Atwood et al., 1990) and Mexican Jays (*Aphelocoma ultramarina*) (Brown and Brown, 1990), both of which also exhibit delayed breeding. Recently, adult survival of American Crows (*Corvus brachyrhynchos*) was found to exceed 90% near Los Angeles, where first-year birds do not appear to breed (C. Caffrey, pers. comm., 1992). No corvid in the world is known to have adult mortality anywhere near as high as that exhibited by 'Alala during the 1970s and early 1980s away from the McCandless Ranch.

Box Table 1: Banding records and resightings

		YEARS																		
Location	**band no.**	**73**	**74**	**75**	**76**	**77**	**78**	**79**	**80**	**81**	**82**	**83**	**84**	**85**	**86**	**87**	**88**	**89**	**90**	**91**
Honaunau	032						*	-	-	B?					#					
	036							*							#					
	042										*				#					
	043										*				#					
	831								AB	B	B				#					
	832								A						#					
	833								*						#					
	834								AB	B	B				#					
	835								*	-	-				#					
Hualalai	025					*	-	B	-	B	B	B	-	-	-	-	-	-	-	-
	002	*	-	-	-										#					
	006		*	-	-										#					
	011			*											#					
	012			*	-	-									#					
	031						*								#					
	836								*						#					
	837									*					#					
	838								A	-					#					
	840								AB						#					
McCandless	019					*	-	-	-	-	-	-	-	-	-	-	-			
	020					*	-	B												
	015			*	-	B?	B	-	-	-	-	-	-	-	-	-	-	-	-	
	035							*												
	038							*	#											
	037							A												
	039							A												
	040							*												

Notes:

1. Band numbers are the last three digits of the eight-digit U. S. Fish and Wildlife Service band.

2. Key to symbols: *: bird born in year; A: bird known to be adult in year; B: bird known to have bred in year; -: bird known to be alive in year; #: bird known to be dead in year. Column of # for 1986 at Honaunau and Hualalai reflects counts in these areas that showed only one bird alive (band number 025).

Data from Banko and Banko, 1980; Giffin, 1991.

Box Figure 1.

On the basis of banding records, it appears reasonable to group the 'Alala into three age groups: juveniles (0-1 year old), yearlings (1-2 years old), and breeding adults (≥2 years old). According to census data, it seems necessary to separate the populations on the McCandless Ranch, which have remained constant, from those elsewhere, which have declined. Nesting information (from Banko and Banko, 1980) plus banding information (supplied by Giffin, 1991) allow estimation of the relevant demographic parameters:

- The number of fledglings produced per adult (half the number of fledglings produced per nest).
- Juvenile survivorship (fledgling to 1 year old), yearling survivorship (1 year old to 2 years old), and adult survivorship (from year t to year t+1, where t ≥2. The numbers shown in Box Table 2 come from calculations at the end of this box.

To predict the likely deterministic dynamics of these populations, let X_t, Y_t and Z_t be the numbers of juveniles, yearlings, and adults, respectively. The equations describing the McCandless situation above are:

$$X_t = 0.46Z_t$$
$$Y_{t+1} = 0.5X_t$$
$$Z_{t+1} = 1.0Y_t + 0.8Z_t$$

For the equations at Hualalai and Honaunau, comparable coefficients can be substituted. These equations can be investigated analytically, but for simplicity we show the results of two simulations below. As can be seen in Box Figure 2, barring unforeseen demographic accidents or chance environmental disturbances, the McCandless population is predicted to show a very slow increase in population size; the populations elsewhere are predicted to decline rapidly (although not as rapidly as actually observed.)

The data and their interpretation follow in Box Table 2.

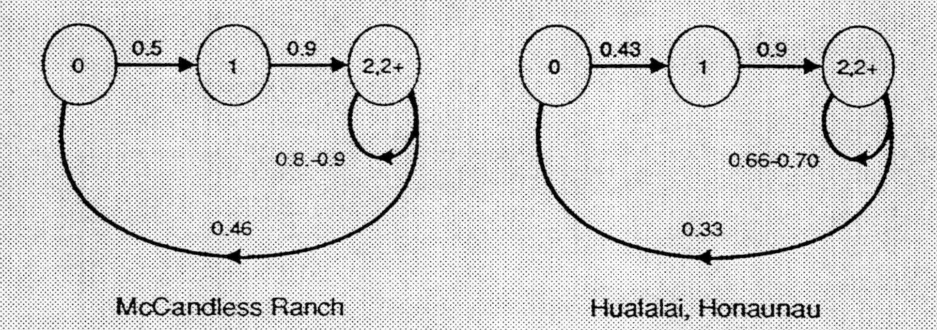

Box Table 2: Analysis of banding records and survival

Juveniles are ≤1 year old, yearlings 1-2 years old, and adults ≥2 years old.

Juvenile Survival: McCandless Ranch

Juvenile survivorship: 3 of 6 survived the first year = 0.50

Juvenile Survival: Honaunau and Hualalai

Juvenile survivorship: 6 of 14 survived the first year = 0.43

Yearling Survival: all sites

Yearling survivorship: 6 of 6 survived the year at Honaunau and Hualalai, and 3 of 3 survived the year at McCandless, resulting in a survivorship of 1.00 at both places. As explained in the text, such a value is not unexpected. However, given that 100% survival is unlikely, we have suggested that it might be lower. A survivorship value of 0.9 is used in Box Figure 1.

Adult survival: McCandless Ranch

Band number	019	020	015	037	039	Total
Number of year-to-year transitions survived	9	0	13	0	0	22
Number of year-to-year transitions possible	10	1	14	1	1	27

Adult survivorship: 22/27 = 0.81. As indicated in the text, this is a minimum value, although, given the rapid disappearance of birds from these sites, it cannot be far below the actual value. Using the time-weighted method (see text), we estimated the adult survivorship as 0.9.

Adult survival: Honaunau and Hualalai

Band number	025	002	006	012	032	831	832	834	835	838	840	Total
Number of year-to-year transitions survived	12	1	0	0	1	2	0	2	0	1	0	19
Number of year-to-year transitions possible	12	2	1	1	2	3	1	3	1	2	1	29

Adult survivorship: 19/29 = 0.66. Again, this is a minimum value, and the time-weighted method yields a value of 0.7.

Box Table 2 [continued]

Analysis of nesting data

Nesting data are drawn from Banko and Banko (1980), who calculated a fledgling rate of about 0.8 bird per pair per year. Three important, related ideas emerge from their raw data:

1. The value of 0.8 is the average value that combines all nests found, including renesting and nests that were found very late in the nesting cycle. Nests that failed before being encountered were not included in any way. (There is a statistical procedure for estimating true nest success from such data with the observed daily probabilities of failure. We did not do this calculation, because we felt that it would not alter the results by very much.)

Renests should not be counted separately in calculating the observed production of fledglings per year. Box Table 4 shows that from 1970 to 1978 in all areas, including McCandless Ranch, 58 fledglings were produced by 67 pairs that attempted to nest. This yields a value of 0.87 fledgling per pair (0.43 per bird).

2. Of more concern is that numerous records exist during the 1970s for which no nesting was documented, despite considerable searching efforts. It is vital to correct our estimates to incorporate these unsuccessful pairs, because those which *did* produce fledglings in unfound nests were incorporated into the overall summary (as "unobserved nests") in Banko and Banko (1980). The pairs that never nested or whose unsuccessful nests were never found are not counted anywhere. We catalogued these pairs as closely as possible with the documentation in Appendix I of Banko and Banko (1980), aided by the narrative on pages 10-25 (see Box Table 4).

Our initial estimate of 0.87 fledgling per pair is therefore almost certainly too high. Virtually all the nonreproductive pairs were encountered at sites where the ʻAlala has been known to breed. In some cases, the observers missed a year, and nests were located at the same sites they were in both the previous and subsequent years. Many represented the final year of documented territorial occupancy before disappearance. We were conservative in counting only the birds and places where permanent occupancy was strongly suspected. Had these pairs actually produced a fledgling in those years, they would have been logged as reproductive and listed in Banko and Banko's (1980) Table 1.

As indicated in Box Table 4, the corrected fledgling production is 58 fledgling in 83 pair-years, or productivity of 0.70 fledgling per pair. That value should be considered the overall average productivity of the ʻAlala islandwide. It is on the low side for the family of Corvidae (see Table 2.4), although data on other species are few.

3. Perhaps most important is that, on the McCandless Ranch, all pairs attempted to breed each year. From 1972 through 1978 on the McCandless Ranch, there were 11 fledglings during 12 pair-years—an average productivity of 0.92 per pair. If those fledglings are subtracted from the total islandwide production, average production for the areas off the ranch was 47 fledglings in 71 pair-years, or 0.66 fledgling per pair.

In sum, we estimate the average fledgling production per individual adult per year to be 0.66/2 = 0.33 in Hualalai and Honaunau and 0.92/2 = 0.46 on the McCandless Ranch. Those are the values found in Box Figure 1.

Box Table 3: Nonreproductive pairs on territories in Kona District, 1971-1978

Ranch	Year							
	1971	1972	1973	1974	1975	1976	1977	1978
Pu'uwa'awa'a	–	–	–	1	1	1-2	–	1?
Hualalai	1	1	1	–	1	1	–	–
Palani	–	1	–	–	1	1	–	–
Honaunau	–	–	–	–	1	–	–	–
McCandless	–	–	–	–	–	–	–	–
Yee Hop	1	–	–	1	–	1	–	–
TOTAL	**2**	**2**	**1**	**2**	**4**	**4-5**	**0**	**1**

Data from Banko and Banko (1980).

Box Table 4: Fledgling production on territories in Kona District, 1970-1978

All areas, including McCandless Ranch

Year	Reproductive Pairs Found	Total Fledglings	Nonreproductive Pairs on Territory	Pairs on McCandless Ranch	Fledglings on McCandless Ranch
1970	3	2	0	–	–
1971	1	0	2	–	–
1972	5	1	2	1	0
1973	9	8	1	1	2
1974	12	11	2	1	0
1975	8	10	4	2	3
1976	11	11	4-5	2	2
1977	9	9	0	2	3
1978	9	6	1	3	1
TOTAL	**67**	**58**	**16 (17)**	**12**	**11**

Fledglings per nesting pair = 0.87
Fledglings per territorial pair = 0.70

Fledglings per territorial pair, off McCandless Ranch (n = 55 pair-years) = 0.66
Fledglings per territorial pair, on McCandless Ranch (n = 12 pair-years) = 0.92

Data from Banko and Banko (1980).

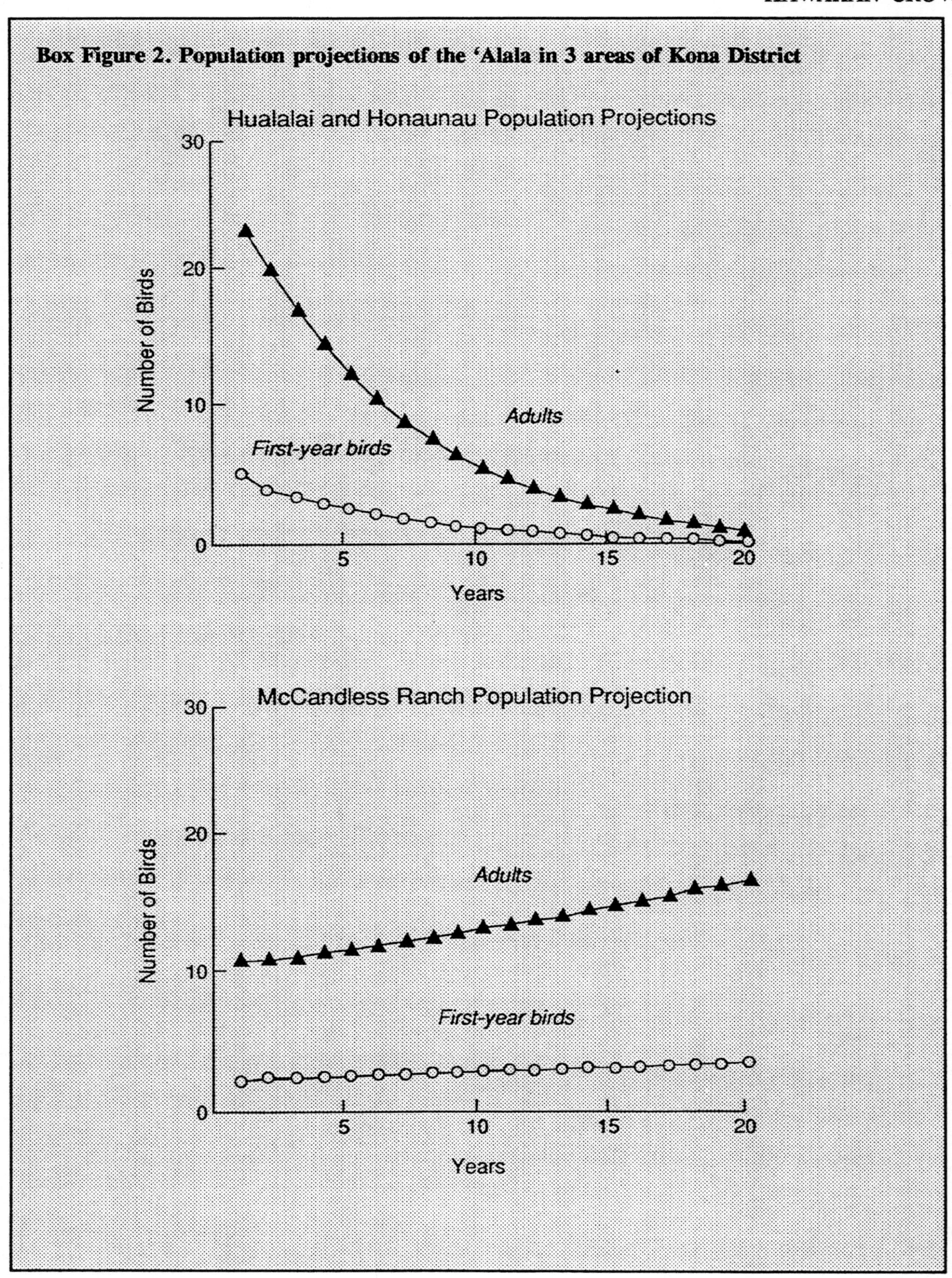

Box Figure 2. Population projections of the 'Alala in 3 areas of Kona District

It is possible that reduced reproductive success, and especially the high frequency of failure of pairs to attempt nesting at Hualalai and at Honaunau, can be attributed to high adult mortality. Especially high mortality among established breeders has the effect of scrambling the few remaining territory holders each year, forcing re-pairing among unfamiliar and inexperienced birds (Woolfenden and Fitzpatrick, 1984, 1991). In a stable population of sedentary corvids, mateless territory holders are quickly discovered by dispersing prebreeders or other mateless breeders and re-pairing is rapid. In fragmented populations, however, solo birds, especially orphaned nonbreeders, might leave the territory in search of other occupied territories and conspecifics. Such a scenario among ʻAlala is supported by the frequent observations of wandering, extra-limital birds during the late 1970s (Banko and Banko, 1980).

It is not clear why adult survivorship on the McCandless Ranch has remained high or why it was lower elsewhere. It should be noted that these estimates of adult survival are strongly influenced by the one or two longest-lived individuals at each location (Box Table 2). Such sensitivity calls attention to the uncertainty inherent in estimating this life-history parameter, and it also suggests the vulnerability of these populations to demographic accidents or chance events. Possibly, by chance alone, a few birds with enough "know-how" (or immunity from disease) to escape the most common causes of death elsewhere are on the ranch. One such bird persisted on Hualalai (025) for at least 12 years, while an average of one-third of her compatriates disappeared *each year*. Such a possibility, of course, is a worst-case scenario: unless their numbers are augmented soon, the death of such special, experienced birds would be accompanied by the final extinction of ʻAlala in the wild.

Population projections based on our estimates of reproduction and first-year, second-year, and adult survival replicate the observed decline (effective extirpation) of non-McCandless ʻAlala (Box Figure 2). Our analyis suggests that the wild population of the ʻAlala on the ranch could be stable or even increasing very slightly. The McCandless Ranch could still be a "net exporter" of ʻAlala, but any dispersing bird would have to colonize areas where high adult mortality is known to have occurred. The analysis also allows us to predict that, even if it is at equilibrium or exporting a few ʻAlala, the McCandless Ranch population will remain at low numbers for decades (Box Figure 2) and cannot exceed a few territories (Figure 2.6), because the social system and ecological requirements of ʻAlala prevent it. Given the low numbers, demographic accidents alone could easily extirpate all remaining ʻAlala in just a few years, and the risk of chance extinction will remain high for decades.

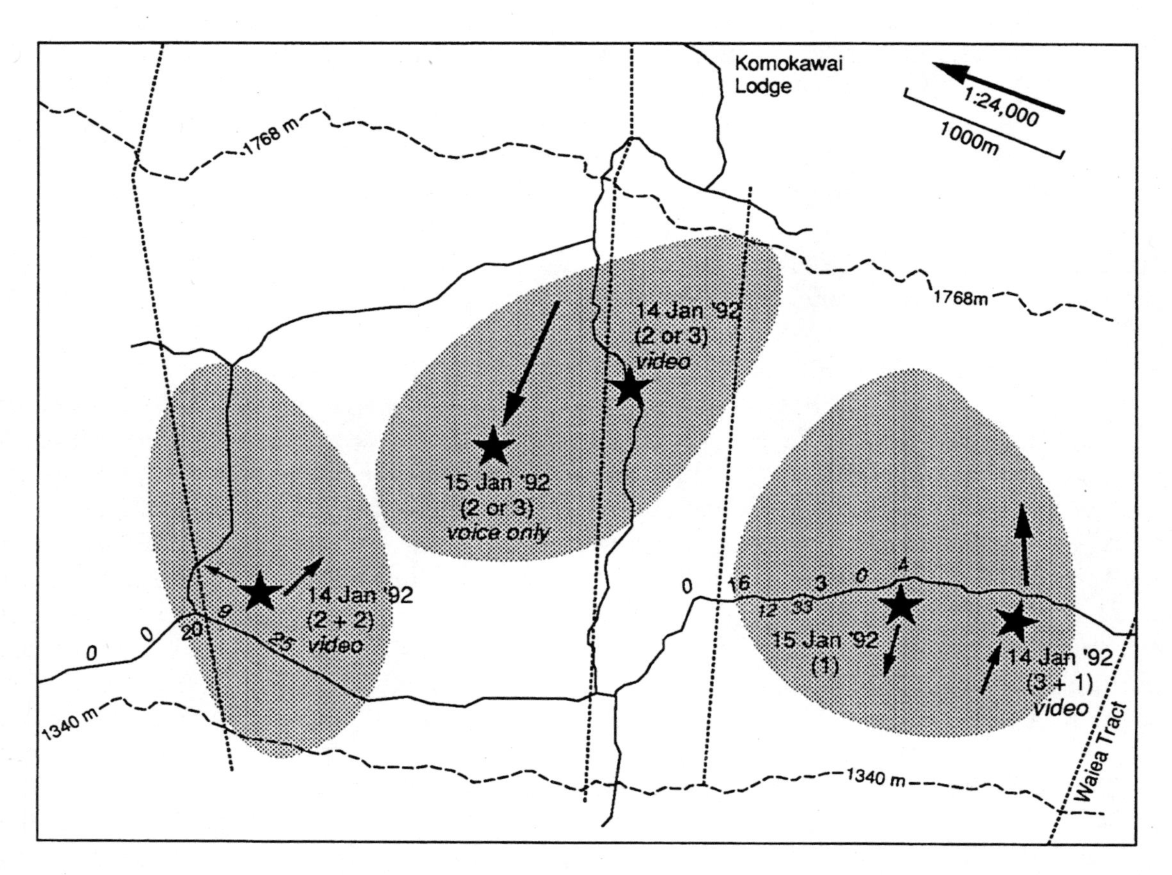

Figure 2.6 Map of the portion of McCandless Ranch where ʻAlala are regularly recorded, according to ranch personnel. Thin solid lines are jeep trails; thin dotted lines are fences (these and contour lines traced from U.S. Geological Survey quadrangle ʻKaunene'). Small numbers along jeep trails indicate numbers of ʻAlala encounters recorded by McCandless Ranch trapper throughout 1991 (N=128 encountered by sight or sound). Trapper noted 1 (N=82), 2 (N=36), 3 (N=10), or 4 (N=3) ʻAlala in each encounter--never more than 4 at one site--and committee had a maximum of 10 total ʻAlala records on 14-15 January 1992. Thick solid lines encircle three hypothetical ʻAlala territories consistent with all observations by committee and ranch personnel. No evidence of more than three territories on McCandless Ranch during the 1970s or 1980s exists. It is possible that only two territories exist and that records suggesting the middle territory (including McCandless personnel records at sandalwood exclosure downslope from Komokawai Lodge) represent birds from the other two territories.

Times to Extinction of Corvids on Islands

There are two broad approaches to estimating time to extinction. The first is used when a population is declining continuously and involves a simple extrapolation to the point of extinction. For instance, extrapolation of 1974-1979 data on the population of 'Alala on Hualalai would have predicted the population's extinction by 1982 or 1983 (see Figure 2.3.a), and indeed 1983 was the last year in which the species bred in that area. Such an approach is not applicable to the small population on the McCandless Ranch, where, in contrast with its behavior everywhere else, the species has not obviously declined in the last decade.

The second approach to predicting time to extinction asks what determines the time to extinction of such small populations that might appear to be stable; it is the approach developed in this section. Two kinds of evidence are examined: from other very small populations of corvid species that have been counted for long periods and from an examination of the known features of the 'Alala's demography (analyzed above).

First, however, we review some of the background on predicting times to extinction. Theoretical studies recognize two kinds of extinction processes that concern small populations. The first, demographic accidents, stem from the vagaries of birth and death. For example, a population consisting of only a few breeding pairs has a high risk of demographic accidents: all the young born in one generation might be of the same sex, or all the adults might die in one generation for independent reasons. Demographic risks quickly diminish as the population increases. In nature, an additional risk comes from events that are external to the population. Unpredictable environmental disturbances can cause many animals to die--a forest fire, a hot summer, a cold winter, an unusual influx of predators, die-off of prey, a hurricane, and so on. Experience with small animal and plant populations shows that small populations, even if they are not declining, are extremely vulnerable to extinction caused by such chance events. The resulting fluctuations in population size impose a considerable risk of extinction, especially for populations that have high year-to-year variations in numbers. Consideration of the magnitude of size fluctuations and the additional risk that they impose can be an important issue in deciding on a management strategy, as they frequently represent the most important factor in determining extinction probabilities (Goodman, 1987).

In addition to the theoretical models, empirical methods can be used to predict how long small populations are likely to last before they become extinct. In the present case, data are available on very small populations of corvid species elsewhere in the world. In addition, other kinds of information might be available on the species of interest:

- *Population data.* How many, where, and how these change over time.
- *Distributional data.* Whether the species is only in habitats larger than some specified size.
- *Life history data.* Age- or stage-specific fecundity and survival.

Even if a great deal is known about a species, we might not know the right things. What is known--and particularly what is not known--dictates the analytical methods that can be used. For example, all that is known about some warblers in the eastern United States is that they are typically absent from woodlots of a particular size--that allows prediction of population if the patterns of forest fragmentation are known. For other species, information about life-history characteristics might permit age-dependent or stage-dependent modeling. These modeling approaches have some severe drawbacks: To be compelling, the predictions need both estimates of parameters and estimates of their variability. For example, for a long-lived species like the 'Alala, stage-dependent models often find adult longevity to be the critical parameter in determining time to extinction. The models merely predict that the population will last until the adults die in some unexpected, episodic way--e.g., as a result of a bad storm, a cold winter, a major forest fire, a drought, or shooting. Without knowledge of the nature and frequency of such deaths, predictions of that type are of limited use.

The second kind of evidence we examine involves estimates of times to extinction for the 'Alala derived from observed times to extinction of very small populations of other species of corvids. There is an extensive theoretical literature on times to extinction, but few data. Entire populations must be counted repeatedly for long periods, and analyses require a large set of such counts. The best estimates of how long small populations of corvids survive come from yearly counts made on islands off the coasts of Great Britain and Ireland. Analysis of those data leads to the conclusion that the median time to extinction for a maximum density of three breeding pairs is 8 years. (See Appendix A for a detailed summary of the derivation of estimates of time to extinction for those small island populations.)

The effective population size of the 'Alala on the McCandless Ranch is likely to be three or fewer breeding pairs. There is the estimate of "3 young produced in the wild in 1991" provided by Peter Simmons, manager of the McCandless Ranch. A maximum of three pairs is also consistent with the observations of this committee in January 1992, and 3 nests were sighted in April 1992 (J. Engring, pers. comm, 1992). A comparison with the island data noted in Appendix A suggests that as long as the 'Alala population on the McCandless Ranch remains at its current size (and does not increase), it has a 50% chance of extinction within a decade (and each subsequent decade).

Suppose that the 'Alala on the McCandless Ranch could be considered a "healthy" small population, i.e., the population is not suffering an inexorable decline and has numbers that remain approximately constant from year to year. Assume that habitat loss and predation are no longer factors, because of the protection afforded the species on the McCandless Ranch; that there is no decline attributable to disease; and so on. In short, the population has now stabilized. The British island populations, too, are "healthy" in this sense. Long-term population data from the British islands show that all but one of the species of crows are increasing; the exception is the Chough, which suffered local declines on Great Britain (now reversed), but tended to be unchanged on the small islands. Despite that favorable news, even "healthy" populations become extinct because of chance events. So the British island data are comparable with the best-case scenario for the 'Alala.

HISTORY OF THE WILD POPULATION AND CAUSES OF ITS DECLINE

Although comparisons with other tropical island endemic species would be valuable, there are few precise data for corvids. Craig (1991) has called attention to the persistence over many decades of small populations (fewer than 20 pairs) of island birds in New Zealand, and Jones and Owadally (1988) summarize examples on Mauritius. On the basis of current knowledge, however, the British crows seem to be the best available models for the 'Alala for estimating the likelihood of extinction resulting from demographic accidents. The likelihood of chance extinction is known to depend generally on the dispersal abilities of the species, how long individuals live, how many young the species produces per year, the magnitude of year-to-year changes in the population, the social behavior of the species, and most important, the first-observed size of the population. The 'Alala population on the McCandless Ranch is isolated, whereas the island populations noted above are not isolated. Indeed, one reason why some of the British island populations last so long might be that the populations received immigrants from the mainland. The immigrants might also rescue a population genetically--providing an influx of genes that helps to overcome the dangers of inbreeding. (A more thorough discussion of inbreeding is found in Chapter 3.) That difference in isolation makes the estimates of how long the populations last very optimistic--truly isolated populations might not last as long.

Corvids, in general, live a long time (see records provided by Clapp et al., 1983), and these British island populations have some of the lowest year-to-year variations in numbers ever observed in any species anywhere (Pimm, 1991). Moreover, tropical birds tend to live a long time, to reproduce slowly, and (on the basis of very few data) supposedly to vary less from year to year than temperate species. Long life and low variability reduce the chance of extinction; slow reproduction increases it. In short, there is nothing obviously special about the basic dynamics of the 'Alala versus other corvids. Therefore, we think that it is unlikely that the island data suggest shorter times to extinction than will be observed in the 'Alala.

Social factors might be very important. Some species of corvids nest colonially. Those species are highly prone to extinction, presumably because a small number of birds cannot survive: the Jackdaw seems to be an example. Were the 'Alala to be colonial or loosely colonial in its nesting habits, the predicted times to extinction would have to be revised sharply downward. (Little or nothing is known about the degree of coloniality of the 'Alala.)

Simply, the 'Alala population has a short time to extinction, because of its size and the inevitable demographic accidents that affect it and other tiny populations. Any additional unexpected environmental disturbance would shorten its time to extinction considerably.

Regardless of the particular causes of the decline of the 'Alala, its management, given its present critical status, should involve all ways to increase its rates of reproduction and survival at all stages of its life history.

The best-case scenario is that the small wild population on the McCandless Ranch is not declining and that three to five pairs of birds are breeding each year. Under such circumstances, normal populations of corvid species--those not undergoing inexorable decline--often encounter local extinction on a scale of 1-2 decades. There is no evidence that a comparable global time to extinction is unrealistic for the 'Alala. Moreover, various factors--such as an inexorable decline because of disease, predation, or a dysfunctional social structure for the species--could all cause the expected times to extinction to be even shorter.

The demography of the population is unusual. Everywhere (including the McCandless Ranch), breeding success, fledgling survival, and yearling survival are not dramatically atypical of corvid species; nor, on the ranch, is adult survivorship. Off the McCandless Ranch, however, *perhaps one-third of all the adults died each year, until the populations became extinct in those areas. That represents a catastrophic annual mortality for such a typically long-lived corvid species.*

Estimates of times to extinction suggest that in the absence of any intervention, the McCandless Ranch population is likely to become extinct within a few decades. Of course, serious environmental disturbance or disease could extirpate a population even much larger than this one.

Population projections based on these admittedly limited demographic parameters suggest that, although the McCandless Ranch population appears stable and may even export a few dispersers, the population will remain at levels at which demographic accidents alone will continue to be a highly probable cause of extinction for the next several decades. For this reason, we believe it unlikely that the 'Alala can survive in the wild without intensive management.

3

GENETIC CONSIDERATIONS

Population size and environment are obviously essential to survival of a species. Less obvious are genetic factors that contribute to survival and reproduction of individuals, and thus species, in nature. Modern population genetics has demonstrated that genes have a strong influence on almost all important characteristics that help a species to adapt to its environment. Such inborn inherited influences reach into every aspect of the life cycle.

Of crucial importance is the great genetic variability between individuals in a population. Individuals show genetic variations that affect not only their overall size, but also the sizes, shapes, and functioning of particular organs. For example, genetic differences may occur in the ability of animals to obtain mates and thus reproduce successfully, in the acuteness of sensory perceptions that will affect their efficiency in finding food and in detecting and escaping from predators, and in such fundamental attributes as the immune system that enables them to withstand attacks of disease.

Breeding in nature normally provides a species with a population consisting of an array of genetically different individuals. Through natural selection, the gene combinations that work well for the species are identified; the hardiest individuals among them leave more offspring. That results in a system for perpetuating and multiplying the "good" variations. Natural selection is not a process that merely eliminates the "bad."

The system is not mysterious. Indeed, practical animal and plant breeders used the same system to select animals and plants long before modern knowledge of genetics emerged. The system essentially retains considerable minor gene variability from which it generates and puts out for trial new variants of the gene combinations that have worked well in the past. It keeps the population large and variable: when there are a lot of combinations (i.e., individuals) to choose from, selective adaptation can keep up genetically with the ebb and flow of environmental shifts.

Even though it works well for most animals and plants, the natural system of selective breeding nevertheless extracts a price from the population, in that it only rarely solidifies and fixes a particular invariable or "ideal" combination of genes. Rather, it retains a balance of many present and potential combinations of genes. The outcome is that some individuals in the population are inferior to others in how they are able to react to the problems posed by their environment.

A further, biologically expensive genetic price is that DNA is prone to pick up and carry along some new mutant genes. Those genes might affect an essential organ system or process negatively; they can thus have a negative effect on the organisms that carry them. Such deleterious genes commonly exist in heterozygous form. When genetic recombination occurs as two carrier parents are producing offspring, the probability increases that one or more of the progeny will get a double dose of (i.e., become homozygous for) one of these "lethal" or "semilethal" genes. The effect is likely to manifest itself by causing the early death of the organism in the embryo stage. If it does not actually kill, it might seriously handicap the organism (a semilethal gene). The existence of such genes in many organisms has been well established. They constitute the notorious "inborn errors," genetic conditions often referred to as our "load of mutations." They have also been studied in experimental genetics of mice and insects, all of which carry genetic diseases in their populations. Many occur naturally, having been hidden in the genome for many generations, but some are caused in current generations by ionizing radiation or mutagenic chemicals.

When populations are large, genetic variation is maintained by selection and mutation, and the general health of the population is not seriously affected when some members die without reproducing or fail to reproduce as well as others. That is the normal situation in the large, natural populations of most species. What happens when the size of the population is reduced, as in an endangered species? Generally, genetic variation is progressively eroded, initially by the loss of rare alleles during the population decline (Denniston, 1977) and later by the reduced effectiveness of selection relative to the chance genetic changes due to the random sampling of alleles in each generation (genetic drift) and the increase in mating between genetically similar relatives (inbreeding) (Frankel and Soulé, 1981; Allendorf and Leary, 1986; Lacy, 1987; Lande, 1988; Simberloff, 1988). The resulting increase in homozygosity leads to increased expression of lethal and semilethal genes that were hidden in the larger population. The situation is well known to animal and plant breeders and to those who deal with small, captive populations of exotic species (see Ralls et al., 1980a). To avoid trouble, breeders must continually strive to keep the stock outbred--to arrange matings between unrelated parents. Indeed, most wild populations have, as part of their inherited breeding systems, many kinds of adaptions that prevent the mating of close relatives; that is, there is a strong tendency for outbreeding, e.g.; sex-biased dispersal in birds and mammals (Greenwood, 1980). Progeny of outbred animals are less likely to manifest the effects of lethal or semilethal genes.

If a population is decreasing in size, generation by generation, its members will be forced by circumstances either to mate with relatives or not to mate at all. That greatly increases the chance of homozygous combinations of lethal or semilethal genes. A common result is a condition called "inbreeding depression"; the term describes the increased difficulty of maintaining genetically healthy animals (Ralls et al., 1980a, b; Ralls and Ballou, 1983; Hedrick and Miller, 1992). It has been dramatically demonstrated in many artificial breeding experiments. For example, using the rapidly breeding, newly domesticated Japanese Quail (*Coturnix japonica*), Sittmann et al. (1966) compared reproductive performance in purposely and Miller, 1992). It has been dramatically demonstrated in many artificial breeding experiments. For example, using the rapidly breeding, newly domesticated Japanese Quail (*Coturnix japonica*), Sittmann et al. (1966) compared reproductive performance in purposely inbred lines and normally outbred lines and determined that successive full-sibling matings led to considerable inbreeding depression in all traits that were considered. Fitness traits--including fertility, egg hatchability, chick survivorship, age at sexual maturity, and total egg production--were affected most. For each 10% increase in inbreeding, fertility declined by 11% (of which 4% resulted from complete male infertility), egg hatchability declined by 7%, chick mortality at 0-5 weeks of age increased by 2-4% and at 5-16 weeks of age by 0.8%, sexual maturity in females was delayed by about 1 day, and total egg production declined by about 1.5 eggs.

Similar, but variable, inbreeding effects have been demonstrated in other avian species, including various strains of poultry (Shoffner, 1948; Lerner, 1954), captive Budgerigars (*Melopsittacus undulatus*) (Daniell and Murray, 1986), captive Mandarin Ducks (*Aix galericulata*) (Greenwell et al., 1982), captive Hawaiian Geese (*Branta sandvicensis*) (Kear and Berger, 1980), captive Pink Pigeons (*Columba mayeri*) (Jones et al., 1989), and wild Great Tits (*Parus major*) (Bulmer, 1973; Van Noordwijk and Scharloo, 1981). Inbred captive lines are usually difficult to keep and are prone to extinction, unless outcrossed or progressively purged of deleterious recessive traits (Templeton and Read, 1983). Much of the above can be demonstrated in many birds, mammals, lizards, and insects (Ralls et al., 1980 a, b; Schoenwald-Cox et al., 1983; Ralls et al., 1986; see also Soulé, 1987).

For the ‘Alala, the major conclusion arrived at in Chapter 2 is that the number of ‘Alala is so low that, unless it is somehow increased very soon, extinction because of accidental loss is virtually certain. Indeed, extinction by chance events is possible even in the unlikely event that the genetic health of the birds is unaffected by the serious population decline of the last 20 or more years.

The wild population of ‘Alala is now apparently reduced to fewer than 12 birds (J. Engbring, pers. comm., 1992), which are not known to be outside a single geographical area of less than 20 km^2 on the McCandless Ranch in the Kona District. Until 11 wild birds were counted by USFWS personnel in March and April 1992 (J. Engbring, pers. comm., 1992), accurate sighting counts were not been made on the McCandless Ranch since about 1980. The proposed number must be viewed in light of the census records cited by Banko and Banko (1980), which covered the 1960s, with 88 birds, and the 1970s, with 346 birds. The latter number was acknowledged to include repeated observation of the same birds in the transect surveys. Nevertheless, in the 1970s, the probable overall population might have been 10-15 times larger than the present estimates. In the 1970s, the birds were distributed along a 75-km portion of the Kona coast from the Pu‘uwa‘awa‘a Ranch south to Alika Homesteads (see Figure 2.1).

In view of their census and restricted geographic range, the wild 'Alala are very likely to be affected to some degree by inbreeding depression. Such a condition can only be surmised, however, because no reliable data on fertility and viability of eggs and young chicks from the wild are available. Most ornithologists studying the birds since 1970 have avoided visiting nests and disturbing the small number of breeding pairs at this crucial stage of the life cycle. Furthermore, substantial inbreeding is now known to characterize some natural and healthy populations of birds that exhibit sedentary habits and group living (Rowley, 1973).

As indicated earlier, the deleterious effects of inbreeding would probably not be expressed solely in reduced fecundity (cf. Hedrick and Miller, 1992). Inbred birds would be expected also to be inferior in vigor, in resistance to disease, and possibly in pair bonding, nest-building, and courtship. Little is known about mate choice in the 'Alala, but in some other birds reduced choice of mates is known to reduce reproductive success (Kepler, 1978; Burley and Moran, 1979; Klint and Enquist, 1980; Bluhm, 1985; Derrickson and Carpenter, 1987; Yamamoto et al., 1989). It should be recalled that in many instances the census count of animals yields a higher number than what population biologists refer to as the effective population size (N_e), which refers to the number of breeding animals (for a discussion of effective population size see Simberloff, 1988). Data on effective population size of the 'Alala have never been obtained or even estimated. Unfortunately for the conservationist, census counts almost never give the exact size of a population, much less its effective size.

Although not all types of population bottlenecks cause genetic disaster (see Carson, 1990; Hedrick and Miller, 1992), the progressive decline in 'Alala numbers has probably had a serious effect on genetic variation. All the 'Alala, both wild and captive, are derived from small populations that have undergone a succession of population bottlenecks or local extinctions extending over many years. Since about 1970, breeding birds have nested in only a narrow elevational and ecological band along the Kona coast that is at most only 75 km long. This very limited breeding range, at an elevation of about 1,000-1,600 m, has been used by crows only at separated local sites (see Figure 2.1) until recently. Only a single small breeding population, on the McCandless Ranch, appears to survive. This persistent pattern of decline suggests that the historical levels of genetic variation might have been progressively eroded as a result of genetic drift and inbreeding. However, because molecular data on both the historical and current levels of genetic variation in the wild not are available, the corroboration of a loss of genetic diversity in 'Alala is lacking. Furthermore, the decline of wild populations has occurred over a relatively long period, so inbreeding might have progressed at a low enough rate to minimize inbreeding depression (Franklin, 1980; Soulé, 1980; Hedrick and Miller, 1992).

If the captive population, which has been shown to be inbred (see discussion in Chapter 4), is to be improved by outcrossing, specimens from McCandless would have to be used. Also as shown in Chapter 4, the McCandless population is so close to the wild sources of some of the captive birds that it would not be expected to provide more than a minor measure of new genetic variability to the captive population. While augmentation of the captive population with wild-origin stock may benefit the captive population genetically, such augmentation is clearly warranted on demographic grounds alone. Therefore, capture and use of wild birds or eggs *solely* for genetic purposes (i.e., separate from a full-scale management plan for all populations of the species) cannot be recommended. It is clear that the preservation of genetic diversity in both the captive and wild populations will require that *both* populations be increased in size as rapidly as possible, and this is best accomplished by improving the reproductive success of both captive and wild crows (Chapter 6).

4

CAPTIVE BREEDING OF THE 'ALALA

Through research and experimentation, biologists have developed several techniques to preserve endangered species (Temple, 1977b). One such technique--captive propagation--has found increasing acceptance as a valid conservation strategy and has been applied to various vertebrate species within the last decade (Cade, 1988; Conway, 1988). Captive breeding has been used principally to avoid extinction of some species, such as the California Condor (*Gymnogyps californianus*)(Snyder and Snyder, 1989), black-footed ferret (*Mustela nigripes*)(Thorne and Williams, 1988), Guam Rail (*Rallus owstoni*), and Micronesian Kingfisher (*Halcyon c. cinnamomina*)(Witteman et al., 1990). In most instances, however, it has been used to complement conventional field research and management efforts to preserve habitat and restore threatened populations. For example, in the case of the Peregrine Falcon (*Falco peregrinus*), captive breeding and reintroduction enabled the restoration of breeding populations in large portions of the former range at a rate far exceeding that which would have occurred through natural colonization (Cade, 1986a,b, 1988).

When wild populations are small, fragmented, and therefore prone to extinction, captive populations can serve several important roles in recovery efforts (Carpenter and Derrickson, 1981; Derrickson and Snyder, 1992). First, and most important, they can provide a safety net to minimize the probability of extinction while preserving genetic variation through controlled pedigree breeding. Second, if captive breeding can be achieved and sustained, captive populations can provide stock for release to the wild. Such releases can be extremely important in correcting sex-ratio imbalances, infusing genetic variation, introducing new behavior, re-establishing extirpated populations, and establishing new populations in natural or altered habitats. In many instances, breeding biology can be manipulated in captivity so that reproductive output can be increased substantially over that obtainable in the wild. Third, captive populations can facilitate research studies on biological, behavioral, or ecological characteristics that are essential for recovery but that are difficult or impossible to address in the wild, owing to small population sizes or environmental conditions. Finally, captive populations and their associated reintroduction programs can provide a foundation for establishing subsidiary programs aimed at habitat preservation, public education, and conservation training (cf. Durrell and Mallinson, 1987; Kleiman, 1989; Butler 1992).

Origins and Facilities of the 'Alala Captive-Breeding Program

The precipitous decline of the wild 'Alala populations in the late 1960s prompted state and federal biologists to endorse the establishment of a captive-breeding program for the species by removing injured or sick birds from the wild. The immediate objectives of the program, like those of similar programs initiated at about the same time for other endangered species, were to avoid extinction in case wild populations were extirpated, to enable research on behavioral and reproductive requirements, and to produce offspring that could be returned to the wild to augment the existing wild populations. The last of those objectives was emphasized in the 'Alala Recovery Plan developed later (Burr et al., 1982).

Two fledglings exhibiting pox lesions were retrieved from adjacent nesting territories on the Hualalai Ranch in June 1970 and were held at the research aviary in Hawai'i Volcanoes National Park. Both were transferred to the endangered-species breeding facilities at the Patuxent Wildlife Research Center in Maryland in August 1970 after treatment and apparent recovery from pox. One of the two died soon after arrival at Patuxent, and the other died, apparently of circulatory problems, on October 1, 1975 (Banko and Banko, 1980). Three additional wild fledglings were salvaged in 1973; two were found on the ground, and the third was discovered to have a heavy infestation of northern fowl mites. All three were successfully raised at the Hawai'i Volcanoes National Park's research aviary and later transferred to the state of Hawai'i's endangered-species breeding facility at Pohakuloa in early 1976 (Banko and Banko, 1980). Between 1977 and 1981, seven more 'Alala were removed from the wild and transferred to the Pohakuloa facility for captive breeding.

The Pohakuloa facility, on the leeward slope of Mauna Kea at an elevation of about 1,950 m, was established in 1949 for the propagation of the Hawaiian Goose, or Nene (*Branta sandvicensis*). By 1976, when the state acquired responsibility for the 'Alala breeding program from the federal government, two additional endangered species were being propagated at Pohakuloa--the Hawaiian Duck (*Anas wyvilliana*) and the Laysan Duck (*Anas laysanensis*). Although considerable success was achieved with the three waterfowl species, the breeding of 'Alala proved much more difficult, despite periodic consultations with aviculturists and zoo personnel. In 1985, as a result of poor 'Alala reproduction and the recognition that propagation efforts would probably be expanded to include other endangered Hawaiian forest birds, the state asked Stanley Temple, of the University of Wisconsin, to explore the potential for developing a new facility at a more suitable location. He suggested that the 'Alala be moved from Pohakuloa for several reasons, including substandard facilities, personnel and predator problems, prevailing climatic conditions, and the periodic disturbance caused by military training activities at the nearby U.S. Army Pohakuloa Training Facility. After reviewing several alternative sites in the state, Temple recommended that the former Olinda Honor Prison Camp, near the town of Makawao on Maui, be renovated as a new propagation facility, which would have the advantages of pre-existing buildings, suitable climate, and relative isolation.

At an elevation of 1,500 m on the windward slope of Haleakala, the 45-acre site is characterized by a cool, wet climate typical of mid-elevation rain-forest habitat. Federal and state funds and in-kind construction assistance from the U.S. Army were used to renovate the facility. Initial work at Olinda focused on the construction of two large subdivided aviaries for the ‘Alala, the remodeling and renovation of buildings for an interim veterinary clinic and staff quarters, and the installation of water and emergency power equipment.

The ‘Alala pens were completed in July 1986, and the nine ‘Alala in the captive flock at Pohakuloa were transferred to Olinda in late November that year. Site renovation and development have continued since 1986, with the addition of waterfowl breeding and rearing complexes, infectious and noninfectious quarantine facilities, security fencing, and miscellaneous renovations of the existing buildings. All the ‘Alala breeding pens are occupied, and only a single juvenile enclosure is unoccupied. Although up to six additional birds could be accommodated by dividing some of the existing enclosures, those modifications would not be suitable for breeding (F. Duvall, pers. comm., 1991, 1992). Plans are now under way to construct additional ‘Alala and forest-bird pens and to expand the existing Nene breeding complex in 1992.

Three full-time staff are employed at Olinda: a director-aviculturist and two animal technicians. Three additional staff members include an avicultural assistant with a 1-year contractual appointment and two laborers with temporary limited-term appointments. Veterinary services, both emergency and routine, are provided by Ben Okimoto of the Honolulu Zoo on O‘ahu; previously, these services were handled by a local veterinarian, Renata Gassman-Duvall, through a contract with the Hawai‘i Department of Land and Natural Resources.

Staff at both Pohakuloa and Olinda have sought the assistance of a variety of aviculturists and institutions through the years. The closest ties have been with staff at the San Diego Zoo. In September 1990, Cynthia Kuehler, a zoologist with the Zoological Society of San Diego and an advisor to this committee, conducted a review of the ‘Alala breeding program and made a number of husbandry and management recommendations to improve the health and productivity of the captive flock, many of which have been implemented. She also reviewed and summarized the existing records and prepared a studbook for the captive population, which has greatly facilitated the demographic and genetic analyses discussed below. A copy of the studbook, listing hypothetical founders and all living and dead ‘Alala in the captive population, is provided in Appendix B.

In March 1991, laparoscopic examinations were conducted on juvenile and nonreproductive adult ‘Alala at Olinda by San Diego Zoo veterinarian Donald Janssen and animal health technician Kim Williams to confirm the birds’ sexes and evaluate their general health.

Demographics and Genetics

The studbook lists a total of 23 'Alala, including three hypothetical founders, 14 wild-caught birds brought into captivity from 1965 to 1983, and six captive-hatched birds. No wild-caught birds have entered the population since 1983, and the population has remained small because of both adult mortality and low reproduction (see Appendix B). Specific causes of mortality have included shipment stress (one case), pericarditis/splenitis (one), toxemia (one), hemochromatosis (three), yolk peritonitis (two), and unknown causes (one). Six pairings, involving four females and four males, resulted in the production of eggs during the period 1979-1991, but only four pairs, involving three females and three males, have produced offspring (Figure 4.1, Table 4.1, Table 4.2).

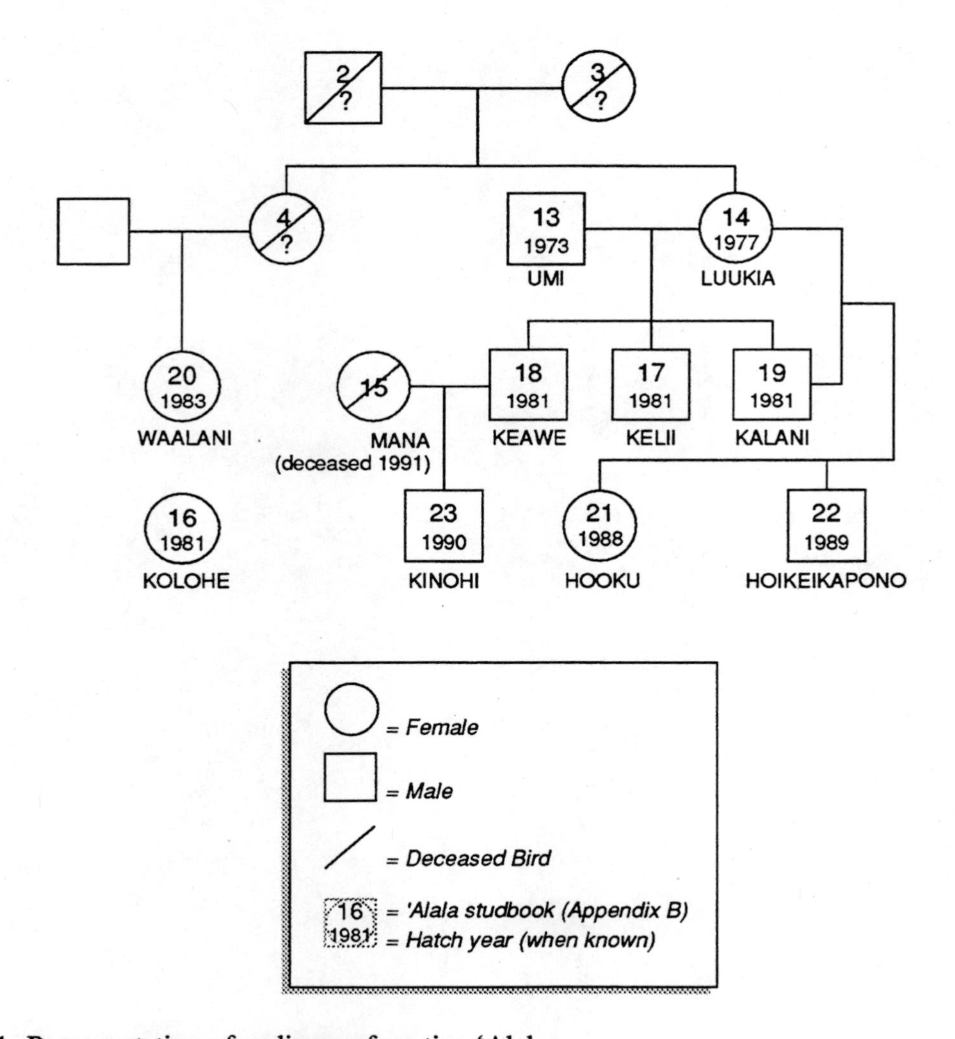

Figure 4.1 Representation of pedigree of captive 'Alala.

Table 4.1 Pairings at Pohakuloa (1979-1986) and Olinda (1987-1991) Facilities

Pairings F	M	Year	No. of eggs	Fledgling survival
Luukia	Umi	1979	3	0
		1980	2	0
		1981	8	3 (Kalani, Keawe, Kelii)
		1986	1	0
Hiialo	Kalani	1984	10	0
		1985	4	0
		1986	2	0
		1987	2	0
		1988	10	1 (Hooku)
Mana	Kelii	1984	2	0
		1985	2	0
Mana	Keawe	1988	3	0
		1988	3 false-lay clutches	0
		1990	5	1 (Kinohi)
		1991	3	0
Luukia	Kalani	1989	8	1 (Hoikei)
		1990	8	0
		1991	2	0
Waalani	Umi	1991	3	0
		Total	81	6

Table 4.2 Summary of 'Alala pairings in the captive flock[a]. In monogamous species of birds, successful reproduction requires physiological synchronization and sexual and social compatibility.

Female	Pairing year(s)	Mate	Comments
Mana	1979-1981	Ulu	Male died July 2, 1981
	1982-1987	Kelii	Productive; pair bond tenuous; pair separated several times, 1984-1987
	1988-1991	Keawe	October 1987; eggs produced 1988, 1990, 1991; female died August 28, 1991
Hiialo	1980-1981	Imia	Male died March 3, 1981
	1982	--	No mate
	1983-1987	Kalani	Productive; female died June 11, 1987 of peritonitis.
Luukia	1979-1987	Umi	Productive 1979-1981; 1982-1986, no nests or eggs; 1986, laid late; 1987, separated because of aggression
	1988	Kalani	Productive pairing
Kolohe [imprinted]	1982-1986	Keawe	Since June 1981 as juvenile; 1982-1986 nests, but no eggs
	1987	Kalani	Incompatible
	1987	Umi	Placed together December; nest built, little sexual interaction; 1991, female avoidance
		Kelii	Placed together June; appear socially and sexually compatible

[a] Information provided by F. Duvall Feb. 3, 1992.

Table 4.2 [continued]

Female	Pairing year(s)	Mate	Comments
Waalani	1987-1991	Kelii	Late nest-building, 1987-1990; male broke egg 1991, no relaying; separated because of aggression
	1991	Umi	Compatible, but nest-building only
Hina	1977-1978	Umi	Productive, chicks and eggs disappeared; female died August 7, 1978, of toxemia
Hooku	1991	Keawe	Placed together in August after death of Mana; appear compatible

ADDITIONAL NOTES REGARDING CHICK REARING:

- Three siblings raised at the Honolulu Zoo were kept together until August 1981, when they were transferred to Pohakuloa facility and placed in separate enclosures at Pohakuloa.

- At Olinda facility young have been raised individually and then placed together as fledglings until approximately 1-year old (when intraspecific aggression occurs); they are then separated and housed individually.

Of the nine birds transferred from Pohakuloa to Olinda in November 1986, six were wild-caught birds originating in three locations along a 75-km stretch of the Kona coast: Hualalai (Hiialo, Luukia, and Waalani), Yee Hop-Alika Homestead (Umi), and Honaunau (Mana and Kolohe). The genetic relationships among these birds are unknown, but it should be noted that all breeding females have been derived from only two areas about 35 km apart, and several from locations adjoining the McCandless Ranch. Although the breeding female Mana died in 1991, she is still genetically represented in the population by a single offspring produced with the male Keawe. The 10-year-old female Kolohe (a potential founder) was behaviorally imprinted on humans at an early age and has not yet reproduced.

CAPTIVE BREEDING OF THE 'ALALA

Of the 10 surviving birds (four females and six males) held at the Olinda facility, six were hatched and raised in captivity (Kelii, Keawe, Kalani, Hooku, Kinohi, and Hoikei), and four were hatched in the wild (Umi, Luukia, Waalani, and Kolohe). As would be expected, the age structure of the captive population is unstable, and the pedigree (Figure 4.1) is relatively shallow.

Because one of the primary goals of endangered-species propagation programs is the conservation of extant genetic variation, the 'Alala pedigree has been analyzed with the Single Population Animal Records Keeping System (SPARKS) studbook package developed by the International Species Inventory System (1990) to calculate founder contributions, inbreeding coefficients, and mean kinship values for the living population. As shown in Figure 4.2, fractional founder contributions within the living population are highly variable. Three of the founders (2, 3, and Umi) are overrepresented, two founders (P4 and Mana) are underrepresented, and one founder (Kolohe) has not contributed at all. Because such unequal founder contributions will lead to lower levels of genetic variation in later generations, current breeding strategies should have as a short-term goal reducing the disparity in founder contributions (cf. Lacy, 1989; Haig et al., 1990; Ballou, 1991) and maximizing the retention of genetic variation. Analysis of the current pedigree suggests that the living population retains only about 75% of the genetic diversity present in the founders, but this would increase to slightly over 88% if Kolohe produced offspring, and possibly even higher yet if additional founders were added to the captive population.

Assuming that all the original founders of the captive population were unrelated, two birds (female Hooku and male Hoikei) have inbreeding coefficients of 0.25, and the remaining birds have inbreeding coefficients of 0. Inbreeding coefficients of potential offspring from all potential and actual pairings in the captive population are shown in Table 4.3. Two of the four current pairs will produce offspring with inbreeding coefficients of 0.25, thereby causing a reduction in overall genetic variation within the population. To manage the population so as to achieve target founder contributions and maximize retained heterozygosity, it would be advisable to pair males and females that have minimum mean kinship values (Table 4.4) (Ballou, 1991). To achieve this goal, pairings on this basis would be: Umi x Waalani, Luukia x Kinohi, Keawe x Hooku, and Kelii x Kolohe. It should be noted that three of these recommended pairings (i.e., Umi x Waalani; Keawe x Hooku; Kelii x Kolohe) are presently being attempted at Olinda (Table 4.4) and thus these male-female pairs appear compatible (Table 4.2). The fourth recommended pairing (i.e., Luukia x Kinohi) has never been tried. Considering that this pairing would require the separation of the breeding pair Luukia x Kelani, and that the upcoming breeding season is fast approaching, the repairing of Luukia with Kinohi should probably be delayed until after the 1992 breeding season.

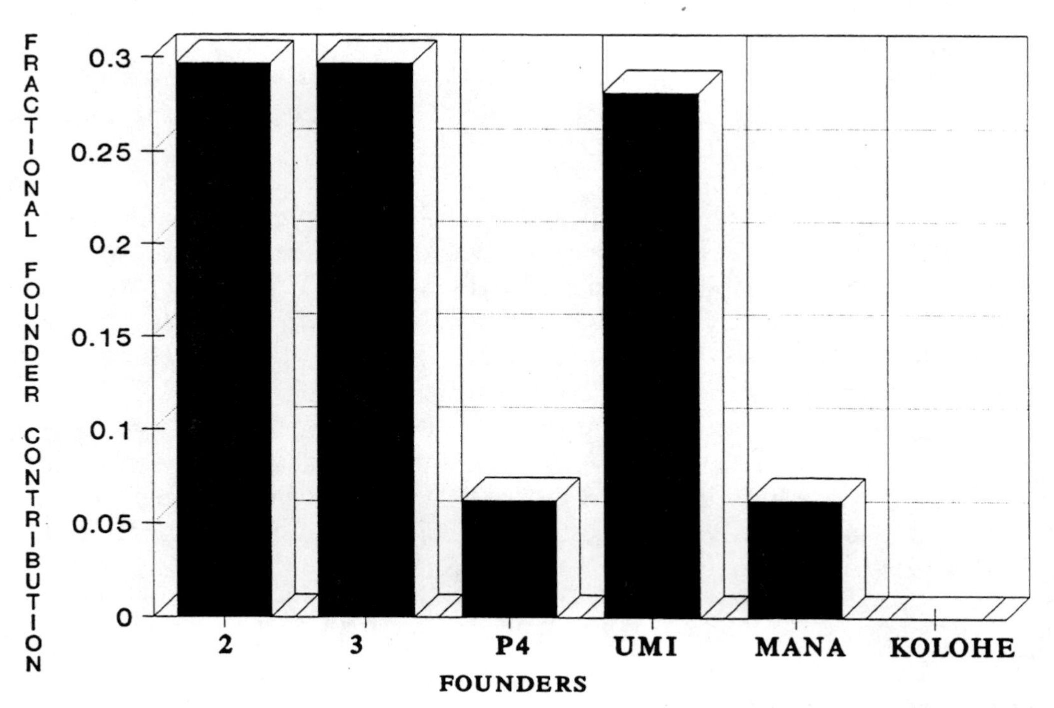

Figure 4.2 Founder Contributions, 'Alala. Studbook numbers beginning with P (see Appendix B) indicate wild or unknown founders that mated with studbook # (without the P) to produce captive-bred offspring.

Table 4.3 Possible pairings and inbreeding coefficients

Female	Male: Umi	Kelii	Keawe	Kalani	Hoikei	Kinohi
Luukia	0.00	0.25	0.25	0.25[a]	0.375	0.125
Kolohe	0.00	0.00[a]	0.00	0.00	0.00	0.00
Waalani	0.00[a]	0.0625	0.0625	0.0625	0.0937	0.0312
Hooku	0.1250	0.250	0.25[a]	0.375	0.375	0.125

NOTE: Males - horizontal
Females - vertical
Current pair. - [a]

Table 4.4 Mean kinship of living birds to living nonfounders.

Name	Sex	Mean Kinship
Umi	M	0.1406
Luukia	F	0.2812
Kolohe	F	0.0000
Kelii	M	0.2422
Kelani	M	0.2734
Keawe	M	0.2578
Hooku	F	0.3086
Hoikei	M	0.3086
Kinohi	M	0.1758
Waalani	F	0.1289

Because the amount of genetic variation that can be retained in a captive population over time is primarily a function of the size of the initial founder population, the rate of population growth to carrying capacity, and the generation length (Soulé et al., 1986; Hedrick and Miller, 1992), it is clear that the captive population needs to be expanded as rapidly as possible. Using the CAPACITY program developed by Jon Ballou of the National Zoological Park, we have examined the population sizes required to maintain extant levels of genetic heterozygosity for various lengths of time. Our results suggest that even when relatively optimistic model parameters are assigned (population growth rate (per year) = 1.5, generation length (years) = 6.0, N_e/N ratio = 0.5, and heterozygosity to date = 88%), the population will have to be expanded to about 130 individuals to retain 75% of the existing heterozygosity over a period of about 50 years. This would essentially require a captive population containing approximately 40 breeding pairs. We believe that this number should be adopted as an interim population objective, while recognizing that this number could potentially be reduced somewhat under certain management strategies, such as the integrated management of both the wild and captive populations (see the discussion of options for management in Chapter 6).

Inbreeding Depression

It has been postulated that the poor success obtained in breeding 'Alala in captivity results from inbreeding depression, and several lines of evidence support this hypothesis. First, as mentioned above, all the captive founders were taken from a relatively small area at a time when the population was declining precipitously; therefore, the assumption that all founders are genetically unrelated might be false. Second, as discussed in Chapter 3, inbreeding is known to affect embryo viability, egg hatchability, and chick survivorship. Between 1979 and 1991, only eight (24%) chicks were hatched from 33 unbroken, fertile eggs, and only six of the eight

(75%) survived. The pattern closely approximates that observed in inbred Pink Pigeons (Jones et al., 1989) and Hawaiian Geese (Kear and Berger, 1980), except that chick survivorship in the 'Alala is considerably higher than that recorded in either of these species. Embryo deaths caused by inbreeding typically occur in the early and later stages of incubation (Romanoff, 1972), and this pattern of mortality has been claimed for the 'Alala. We have reviewed information provided by several sources, but are unable to conclude that such a pattern exists, owing to inaccuracies, discrepancies and omissions in the records, the relatively small sample of eggs, and our inability to rule out other potential sources of embryo death, such as bacterial infections, vitamin deficiencies, and mishandling of eggs (cf. Kuehler and Good, 1990). We do note, however, that the highest embryo mortality has occurred in a pair (Keawe x Mana) whose offspring have the highest inbreeding coefficient (0.25) of all chicks produced to date. It is also interesting to note that this female (Mana) has produced the largest number of abnormal eggs.

Finally, analyses of DNA "fingerprints" of seven captive 'Alala suggest that genetic variation in the captive population is very low, as evidenced by high band-sharing coefficients (Duvall et al., 1991; Fleischer and Tarr, pers. comm., 1992). We do not know the degree to which wild 'Alala would have shared DNA-fingerprint bands when the population was larger, although this is currently being examined by removing DNA from museum specimens (Fleischer and Tarr, pers. comm., 1992). However, several other corvids have been examined recently (Figure 4.3), and the results support a hypothesis of elevated inbreeding among the captive 'Alala. J. Quinn (pers. comm., 1992) used Jeffreys 33.15 probe on a large sample of Florida Scrub Jays and found only 20% bandsharing among unrelated jays. Similarly low band-sharing occurred among unrelated American Crows (Fleischer and Tarr, pers. comm., 1992) and Common Ravens (P. Rabenold, pers. comm, 1992). In contrast, apparently unrelated 'Alala in captivity shared 60% of their bands using both Jeffreys 33.15 and M13 probes (Figure 4.3). This finding reinforces the necessity of both expanding the captive population as rapidly as possible, and adopting breeding strategies that minimize the rate of inbreeding within the captive population.

Brock (1991) recently compared band-sharing in Hispaniolan Parrots (*Amazona ventralis*) and the highly endangered Puerto Rican Parrot (*A. vittata*). She found that the average band-sharing coefficient was higher for unrelated Puerto Rican Parrots than for unrelated Hispaniolan Parrots and that, among mated pairs of captive Puerto Rican Parrots, reproductive success declined as band-sharing coefficients increased.

It must be pointed out that some of the captive 'Alala presumed to be "unrelated" (Figure 4.1) could have shared parents or grandparents. Furthermore, it is possible that small effective population sizes and high variance in lifetime reproduction led to reduced genetic variation (hence high average band-sharing) long before the population declines of 'Alala during this century. A high degree of band-sharing and apparent inbreeding among the captive 'Alala today, therefore, does not necessarily reflect an abnormally elevated level of inbreeding among the remnant wild population of the 1970s and 1980s.

We have no information on whether or not reproduction or survival of the wild birds during those years was affected by inbreeding. We also cannot be certain to what degree the existing level of inbreeding in the captive flock is affecting their ability to survive or reproduce.

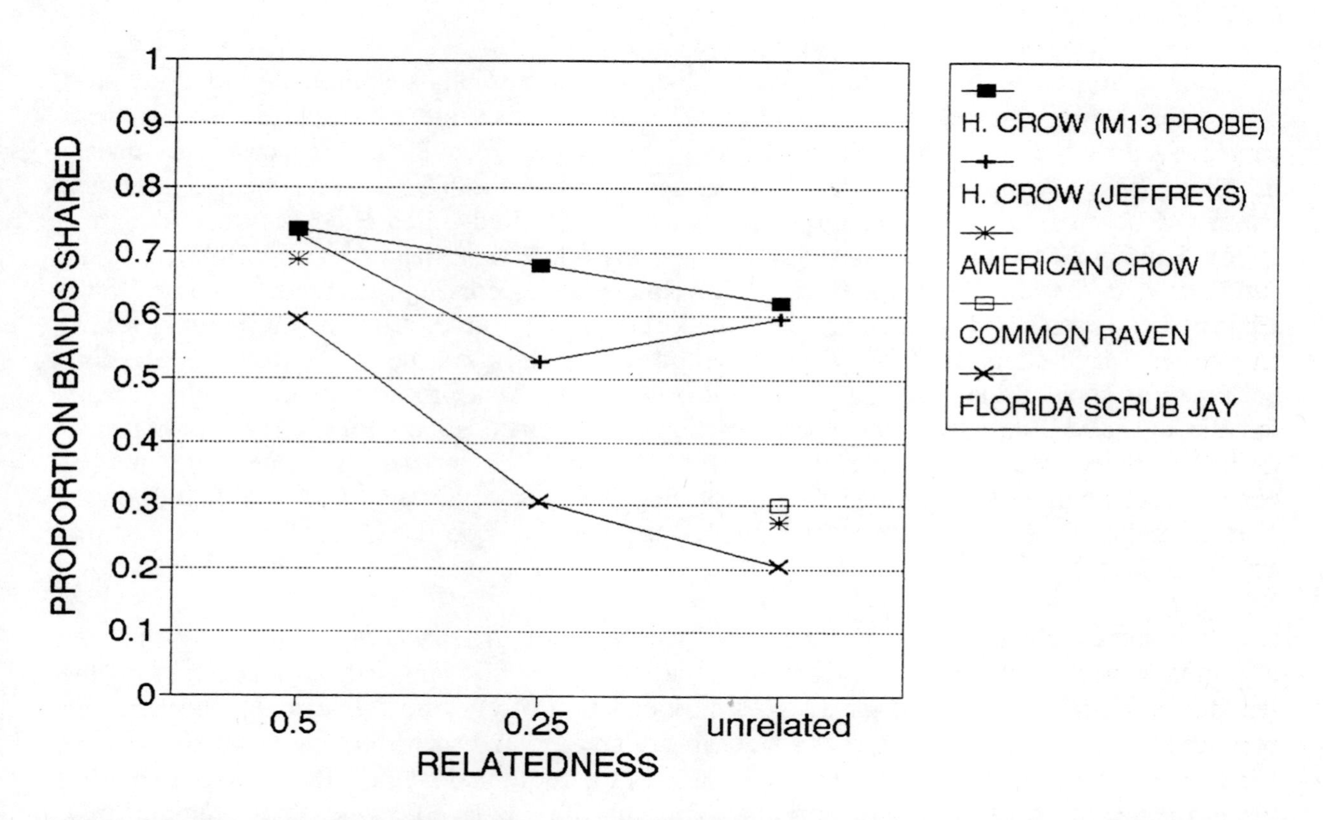

Figure 4.3 Comparison of DNA fingerprint band-sharing coefficients among 7 captive 'Alala (using 2 different probes) and wild populations of three other corvids (all using Jeffreys 33.15 probe). Unpublished Florida Scrub Jay results (N = 16 closely-related, 6 moderately-related, and 56 unrelated pair-wise comparisons) provided to the committee by J. Quinn, J.W. Fitzpatrick, and G.E. Woolfenden (pers. comm., 1992).

Animal Husbandry and Management

The husbandry and management of the captive 'Alala at Olinda are inadequate, and there is room for modification and improvement of existing methods and procedures. There is a strong tendency to become locked into invariable patterns when working with endangered species, and there is always room for experimentation. The old adage "nothing ventured, nothing gained" applies in aviculture as in other vocations, and creativity and experimentation should always be encouraged. Continual review of the avicultural literature and communication with colleagues are essential for developing or applying new techniques and procedures. To date, 39 species of corvids, including nine *Corvus* species, have been bred successfully in captivity (Table 4.5).

Many 'Alala eggs have been broken by the parents, and this source of egg loss has been a substantial drain on potential production. There is also some evidence that chicks were cannibalized by the parents at Pohakuloa. Egg-breaking and cannibalism, especially by males, have long been known in the breeding of captive corvids and have been documented in a wide variety of species, e.g., Eurasian Jay (Delacour, 1936), Red-Billed Blue Magpie (Richards, 1976), Southern Tree Pie (Roots, 1970), Crested Jay (C. Pickett, pers. comm., 1992), Chough (Edworthy, 1972), and Rook (Richards, 1976). In discussing corvid propagation, Shelton (1989) identified egg-eating and cannibalism as an important, but solvable, problem. With two species--the Green Magpie and the Magpie Jay--Shelton simply removed the males from the breeding enclosure as soon as the first egg of the clutch was laid. That procedure works well, because females in most, if not all, corvid species perform all the incubation duties. Also, inasmuch as captive females need only go to a food pan for their own food and that of the nestlings, neither eggs nor nestlings suffer from neglect. Relocating the male out of sight and, if possible, out of hearing range of the female has worked well at the Houston Zoo (L. C. Shelton, pers. comm., 1991, 1992).

The same technique should be attempted with pairs of 'Alala where egg-breaking has been documented and might provide an additional advantage of allowing natural incubation by the female for a period of several days. Natural incubation of eggs for 5-7 days has been shown in a number of species to enhance egg hatchability greatly, and such short-term incubation does not usually interfere with efforts to recycle females for additional laying. Repeated relaying has been documented in a variety of corvids, including the 'Alala. Because the development of effective procedures for full-term artificial incubation will continue to be hampered by the inadequate number of eggs laid each year, initial incubation by the female is strongly recommended. Corvids in most zoos and private collections are allowed to incubate and raise their own young, and with experience they do so efficiently. Consideration should also be given to allowing selected pairs to incubate and hatch their second or third clutches and rear the young. Such reproductive experience can act to strengthen the pair bond and increase future reproductive performance (cf. Derrickson and Carpenter, 1987).

Table 4.5 Captive Breeding of Corvidae

Genus and species [a]	Common Name	No. Successful Breeding Locations, 1961-1991	Multiple Generations	References [b]
Platylophus				
galericulatus	Crested Jay	1-3	No	ISIS
Cyanocitta				
cristata	Blue Jay	4-6	No	IZY
stelleri	Steller's Jay	4-6	No	Risdon (1960); Gibson (1980); ISIS
Aphelocoma				
coerulescens	Scrub Jay	1-3	No	Partridge (1966); Webber and Cox (1987)
Cissilopha				
melanocyanea	Bushy-Crested Jay	1-3	No	IZY
sanblasiana	San Blas Jay	>6	No	IZY; ISIS
beecheii	Purplish-Backed Jay	1-3	Yes	Todd (1980); IZY; ISIS
Cyanocorax				
caeruleus	Azure Jay	1-3	No	IZY
cristatellus	Curl-Crested Jay	1-3	No	IZY
chrysops	Plush-Crested Jay	>6	Yes	IZY; ISIS; Coles (1980)
mysticalis	White-Tailed Jay	1-3	No	ISIS
yncas	Green Jay	>6	Yes	Roles (1971); IZY; ISIS
Calocitta				
formosa	Magpie Jay	4-6	Yes	IZY; ISIS
Garrulus				
glandarius	Eurasian Jay	4-6	No	IZY
lanceolatus	Lanceolated Jay	1-3	No	Goodwin (1954)
lidthii	Purple Jay	1-3	No	Oliver (1964); IZY

Table 4.5 [continued]

Genus and species [a]	Common Name	No. Successful Breeding Locations, 1961-1991	Multiple Generations	References [b]
Urocissa				
caerulea	Taiwan Blue Magpie	4	No	IZY; ISIS
flavirostris	Yellow-Billed Blue Magpie	1-3	No	IZY; ISIS
erythrorhyncha	Red-Billed Blue Magpie	>6	Yes	IZY; ISIS, Richards (1976)
Cissa				
chinensis	Green Magpie	>6	No	IZY; ISIS; Coles (1978)
thalassina	Short-Tailed Magpie	1-3	No	ISIS
Cyanopica				
cyana	Azure-Winged Magpie	>6	Yes	IZY; ISIS; Wayre (1970a)
Dendrocitta				
vagabunda	Rufous Treepie	4-6	No	IZY; ISIS
leucogastra	Southern Treepie	1-3	No	Roots (1970); IZY
frontalis	Collared Treepie	1-3	No	IZY
Crypsirina				
temia	Racket-Tailed Treepie	1-3	No	IZY; ISIS
Pica				
pica	Magpie	>6	Yes	IZY; ISIS
Nucifraga				
caryocatactes	Nutcracker	1-3	Yes	IZY

Table 4.5 [continued]

Genus and species [a]	Common Name	No. Successful Breeding Locations, 1961-1991	Multiple Generations	References [b]
Pyrrhocorax				
pyrrhocorax	Chough	>6	Yes	IZY; ISIS; Edworthy (1972)
graculus	Alpine Chough	>6	Yes	Wayre (1970b); IZY
Corvus				
monedula	Jackdaw	>6	Yes	IZY
frugilegus	Rook	4-6	No	IZY; Richards (1973)
brachyrhynchos	Common (American) Crow	1-3	No	IZY
corone	Carrion (Hooded) Crow	4-6	No	IZY
alba	Pied Crow	1-3	No	IZY; ISIS
hawaiiensis	Hawaiian Crow	1-3	Yes	IZY
ruficollis	Brown-Necked Raven	1-3	No	IZY
corax	Raven	>6	Yes	IZY; Gwinner (1965); Wolinski (1989)
albicollis	White-Necked Raven	1-3	No	ISIS

[a] Taxonomy follows Morony et al. (1975).

[b] IZY, International Zoo Yearbook, Vols. 1-29 (1959-1987); ISIS, International Species Inventory System, Species Distribution Reports 1984-1991; L.C. Shelton (pers. comm., 1991, 1992).

Usnik (1990) suggested that attendants' time spent around the birds during the day during feeding and cleaning of cages should be increased over the normal pattern and noted that this appeared not to affect the birds. We wholeheartedly concur. Increased contact with the birds will condition them to the human presence and in the long-run will make unexpected periods of contact much less stressful. Sanitation in the enclosure would also be improved substantially.

Because 'Alala are territorial, we believe that visual barriers should be used to prevent adjacent pairs from seeing each other directly. In many species, when males can see each other but are thwarted from physical pursuit or contact, females can become the victims of redirected aggression. The use of visual barriers would be best applied initially to pairs that have had a history of sexual or social incompatibility or that have exhibited male aggression directed at the female. Consideration should also be given to increasing the complexity and naturalness of the breeding and juvenile enclosures by providing more plantings, perching material, and nesting baskets. The latter could be placed appropriately to accommodate remote monitoring.

At present, the captive 'Alala receive a rather complicated diet consisting of at least eight food types (e.g., native fruits, commercial exotic fruit, nectar, corn, dry Hi-Pro dog kibble, low-iron bird of paradise pellets, two 1-day-old chicks or a single rat pup, mealworms, and a blended egg "omelet"). This type of diet can pose a number of problems. First, when offered such a variety, many birds will consume only those items that they prefer, and this selectivity can lead to nutritional deficiencies. Such selective feeding has been noted in the captive 'Alala (F. Duvall, pers. comm., 1991, 1992) although its nutritional impact remains unknown. Second, dominant males can monopolize or consume items essential for the female. Third, a diet that includes chicks and rat pups can expose the birds to gram-negative bacteria, which can adversely affect reproduction. Because crows frequently cache food before feeding, the possibility of bacterial contamination is increased.

Our final concern is related to the dietary amounts of heme iron to which the birds can be exposed: three captive 'Alala have succumbed to iron storage disease--Kekau (Patuxent, 1978), Iola-Elau (Pohakuloa, 1979), and Imia (Pohakuloa, 1981). Although the etiology of this disease is poorly understood (Fairbanks et al., 1971; Lowenstein and Petrak, 1978; Davies, 1979; Taylor, 1985; Kincaid and Stoskopf, 1987), it is clear that many fruit-eating species--such as birds of paradise, mynahs, cotingas, and toucans--and others are susceptible to it when maintained on captive diets high in heme iron and acidic fruits. Both chicks and rat pups fed at Olinda are high in heme iron. Serious consideration should be given to simplifying the diet greatly--reducing it to one or two principal items that are palatable, are nutritionally balanced, and contain nonheme iron, and then supplementing this base diet with available native fruits. Use of such a simplified diet would also reduce cost and preparation time.

Staffing and Administration

As a result of interviews with current and former Olinda staff members and the review of internal documents and memoranda provided by USFWS and the Hawai'i Department of Land and Natural Resources, the committee concludes that there have been substantial administrative and personnel problems at Olinda and that staffing and staff training requirements need to be increased. It is clear to the committee that the programmatic problems and deficiencies are recognized by the responsible administrators in these two agencies and that appropriate remedial actions have been initiated or are under review. The committee believes, however, that these problems must be addressed and fully resolved as soon as possible.

The proposed expansion of the Olinda facility to include propagation efforts for three more species of endangered Hawaiian forest birds will obviously require additional staff if the operation is to run effectively 7 days per week. At a minimum, the personnel added to the existing staff should include a facility administrator-curator, a full-time veterinarian, another avicultural assistant-lead keeper, and two animal caretakers.

While the 'Alala facilities at Olinda clearly need to be expanded in the immediate future, breeding facilities should also be developed at an additional, second site (possibly even a third site) on the Hawaiian islands to minimize the possibility of significant losses resulting from disease outbreaks or other potential catastrophes. Because many captive breeding programs for endangered species have suffered severe setbacks as a result of disease outbreaks (see Derrickson and Snyder 1992 for examples and discussion), division of the captive flock and the adoption of strict quarantine procedures and protocols must be considered high priorities.

Additional specific recommendations for the improvement of the captive-breeding program are presented in Chapter 7.

5

RELEVANT PRECEDENTS IN ENDANGERED SPECIES PRESERVATION

Avian extinction during historical times has been largely an island phenomenon. As noted by King (1985), about 93% of all recorded avian extinctions have involved insular species and subspecies and have resulted directly or indirectly from human activity. In light of the findings regarding recent fossil deposits on the Hawaiian islands (Olson and James, 1982a,b, 1991; James and Olson, 1991), and elsewhere in Polynesia (Steadman, 1989), the magnitude of human-caused extinction there is even greater than that noted by King. The increased vulnerability of insular species to extinction can be traced to some combination of their restricted distribution, demographic instability, inbreeding depression, absence of native predators and hence vulnerability to exotic ones, and adverse environmental conditions (Wilcox, 1980; Terborgh and Winter, 1980; Diamond, 1985), but an increasing number of continental species have assumed similar characteristics because of burgeoning human population. The results of continued fragmentation and alteration of natural habitat all over the world can be seen clearly by comparing the threatened corvid species in the International Union for the Conservation of Nature Red Data Book (King, 1977-1979) with more recent compilations (Collar and Andrew, 1988; International Union for the Conservation of Nature, 1990). Although the number of endangered corvid species has remained at two ('Alala and the Marianas Crow, *Corvus kubaryi*), 10 species (three insular and seven continental) have been added to the more recent compilations and are considered at risk (Table 5.1).

The underlying objective of all endangered-species restoration programs is to minimize the probability of extinction by increasing the numbers and the distribution of the species. Such programs typically combine conventional, long-term management activities (protection and restoration of habitat, legal protection, and public education) with manipulative, or "clinical," techniques aimed at immediate enhancement of fecundity and survivorship. Manipulative techniques that have been successfully applied in recovery programs range from such in situ measures as predator control, nest-site provisioning, reproductive manipulation, and supplemental feeding to ex situ measures such as translocation, captive breeding and reintroduction, and cryopreservation of gametes and embryos (Temple, 1977b; Cade, 1986a, 1988; Conway, 1988; Seal, 1988). Some ex situ techniques, particularly captive breeding and reintroduction, remain controversial (cf. Derrickson and Snyder, 1992), although they have been incorporated in recovery programs for a variety of avian taxa (Table 5.2). Many recovery programs have direct relevance to the restoration of the 'Alala, and several relevant cases are described in this chapter.

Table 5.1 Conservation status of corvids[a]

Common Names[b] and scientific	Conservation Status	Geographical Distribution and Reason for Conservation Status
Beautiful Jay *Cyanolyca pulchra*	Insufficiently known	Ecuador and Columbia; rare and local
Dwarf Jay *C. nana*	Insufficiently known	Now restricted to Oaxaca, Mexico; logging
White-Throated Jay *C. mirabilis*	Insufficiently known	Guerrero and Oaxaca, Mexico; habitat logging and grazing
Azure Jay *Cyanocorax caeruleu*	Insufficiently known	Steep declines throughout most of range in Brazil and Argentina
Sichuan Grey Jay *Perisoreus internigrans*	Vulnerable	Pine and coniferous forests of central China
Sri Lanka Magpie *Urocissa ornata**	Vulnerable	Endemic to Sri Lanka; forest degradation
Hooded Treepie *Crypsirina cucullata*	Vulnerable	Central Burma; forest degeneration
Ethiopian Bush-Crow *Zavattariorinis stresemanni*	Rare	Ethiopia; habitat alteration
Banggai Crow *Corvus unicolor**	Rare	Banggai, Indonesia; known from only two specimens; habitat degradation
Flores Crow *C. florensis**	Rare	Flores, Indonesia; habitat degradation; has not adapted to agriculture
Marianas Crow *C. kubaryi**	Endangered	Guam and Rota
Hawaiian Crow *C. hawaiiensis**	Endangered	Hawai'i

[a] From Collar and Andrew, (1988); International Union for the Conservation of Nature Red Data Book, (1990).

[b] *=Insular species.

Table 5.2 Translocations and reintroductions of threatened avian species

Common and Scientific Names; Locality	Methods[a]	References
Yellow-Crowned Night Heron, *Nycticorax violacea;* Bermuda	CR/R	Wingate (1982, 1985)
White Stork, *Ciconia ciconia;* Europe	CB/R	Bloesch (1980), Blackwell (1990)
Atlantic Puffin, *Fratercula arctica;* United States	T	Kress (1977), Cade (1986a)
Bean Goose, *Anser fabalis;* Sweden	CB/CF	Von Essen (1982), Morner (1986)
Lesser White-Fronted Goose, *Anser erythropus;* Sweden	CB/CF	Von Essen (1982), Morner (1986)
Hawaiian Goose, *Branta sandvicensis;* Hawaiian Islands	CB/R	Kear (1986), Kear and Berger (1980), Hoshide et al. (1990)
Aleutian Canada Goose, *Branta canadensis leucopareia;* United States	CB/R;T	Springer et al. (1977), Kear (1986)
Brown Teal, *Anas auklandica chlorotis;* New Zealand	CB/R	Reid and Roderick (1973)
New Zealand Scaup, *Aythya novaesaelandiae;* New Zealand	CB/R	Reid and Roderick (1973)
California Condor[b], *Gymnogyps californianus;* United States	CR,CB/H	Toone and Risser (1988), Wallace (1990, 1991), Kuehler et al. (1991), M.P. Wallace (pers. comm. 1992)

[a] T=translocation of wild eggs, young, or adults; CR=captive rearing of wild-origin stock; CB=captive breeding; R=release of young; H=hacking of young; F=fostering of eggs or young; CF=cross-fostering of eggs or young.

[b] Discussed in the text of the chapter.

Table 5.2 [continued]

Common and Scientific Names; Locality	Methods	References
Andean Condor *Vultur gryphus;* Peru, Colombia, United States	CB/H	Wallace and Temple (1987), USFWS and California Dept. of Fish & Game (1988) Wallace (1990, 1991), Lieberman et al. (1991)
Bald Eagle, *Hailiaeetus leucocephalus;* United States	CR,CB/F,H	Cade (1986a, b), Nye (1988), Laycock (1991)
White-Tailed Sea Eagle, *Hailiaeetus albicilla;* Europe	CB/H	Love (1984), Fentzloff (1984), Cade (1986a)
Bearded Vulture, *Gypaetus barbatus aureus;* Europe	CB/H	Anderegg et al. (1983)
Griffon Vulture, *Gyps fulvus;* Europe	CB/H	Terrasse (1983), Snyder (1986)
Mauritius Kestrel[b], *Falco punctatus;* Mauritius	CR,CB/H,F	Temple (1977a), Jones and Owadally (1988), Jones et al. (1991)
Seychelles Kestrel, *Falco araea;* Praslin Island, Seychelles	T	Cade (1986a), Watson (1989)
Peregrine Falcon, *Falco peregrinus;* United States, Europe	CR,CB/H,F,CF	Cade and Hardaswick (1985), Cade (1986a, 1990), Cade et al. (1988)
Masked Bobwhite, *Colinus virginianus ridgwayi;* United States	CB/F	Ellis et al. (1977), Carpenter et al. (1991)
Cheer Pheasant, *Catreus wallichi;* Pakistan	CB/R	Grahame (1980, 1988), Ridley (1986)
Mississippi Sandhill Crane, *Grus canadensis pulla;* United States	CB/H,F	Zwank and Derrickson (1981), Zwank and Wilson (1988), McMillan et al. (1987)
Whooping Crane, *Grus americana;* United States	CB/CF,H	Drewien and Bizeau (1977), Derrickson (1985) Derrickson and Carpenter (1987), Lewis (1990)

Table 5.2 [continued]

Common and Scientific Names; Locality	Methods	References
Guam Rail, *Rallus owstoni;* Rota, Mariana Islands	CB/R	Witteman et al. (1990)
Weka, *Gallirallus australis greyi;* New Zealand	T,CB/R	Anon. (1991a)
Lord Howe Island Woodhen[b], *Tricholimnas sylvestris;* Lord Howe Island	CB/R	Lourie-Fraser (1983), Miller and Mullette (1985), Fullager (1985)
Takahe[b], *Notornis mantelli;* New Zealand	T;CR/R;CB/R	Reed and Merton (1991), Anon. (1991b)
Black Stilt, *Himantopus novaezelandiae;* New Zealand	F/CF;CB/R	Bryant (1985), Reed and Merton (1991)
Pink Pigeon, *Columba mayeri;* Mauritius	CB/R	Todd (1985), Jones and Owadally (1988), Jones et al. (1988)
Red-Crowned Parakeet *Cyanorhamphus novaezelandiae;* New Zealand	CB/R	Taylor (1985), Wiley et al. (1992)
Military Macaw, *Ara militaris;* Guatemala	CB/R	Clubb (1991)
Thick-Billed Parrot, *Rhynchopsitta pachyrhyncha;* United States	T,CB/R	Snyder and Wallace (1987), Snyder and Johnson (1988), Wiley et al. (1992)
Puerto Rican Parrot, *Amazona vittata;* Puerto Rico	CB/F	Snyder et al. (1987), Wiley et al. (1992)
Kakapo[b], *Strigops habroptilus;* New Zealand	T	Merton and Empson (1989), Triggs et al. (1989), Moorehouse and Powlesland (1991)
Eagle Owl, *Bubo bubo;* Sweden, Germany	CB/H	Broo (1978), Von Frankenburg et al. (1984)

Table 5.2 [continued]

Common and Scientific Names; Locality	Methods	References
San Clemente Loggerhead Shrike, *Lanius ludovicianus mearnsi*; United States	CR/H	Scott and Morrison (1990), Morrison (1991), Kuehler (in press)
Seychelles Brush Warbler, *Acrocephalus sechellensis;* Aride Island, Seychelles	T	Anon. (1988), N.J. Collar (pers. comm., 1990)
Chatham Island Black Robin[b], *Petroica traversi;* New Zealand	T/F,CF	Merton (1975), Flack (1977), Reed and Merton (1991)
White-Eye Vireo, *Vireo griseus bermudianus;* Bermuda	T	Wingate (1985)
Laysan Finch, *Telespyza cantans;* Hawaiian Islands	T	Conant (1988)
Bali Mynah, *Leucopsar rothschildi;* Indonesia	CB/R	Van Balen et al. (1990), Anon. (1991c), Quammen (1991), Van Bali, Balen and Soetawidjaya (1991)
Saddleback, *Creadion carunculatus;* New Zealand	T	Merton (1975), Mills and Williams (1979)

California Condor (*Gymnogyps californianus*)

The recovery program for the California Condor is probably one of the most controversial ever undertaken. Although a succession of biologists tracked the condor's decline for over 40 years, the causes remained undetermined. In 1979, the population had dropped to about 30 birds, and a new research program was initiated primarily under the supervision of the U.S. Fish and Wildlife Service (USFWS) and the National Audubon Society. Initial proposals to trap condors for radiotelemetry and captive propagation brought forth a large opposition. Through the effective use of the press and political pressure, opponents sought to stop all intensive research and "hands-on" management activities. The death of a wild nestling during handling in 1980 reinforced their case, and intensive procedures had to be abandoned.

For the next several years, research activities were limited to observation and photography. During that time, it was learned that most of the adult birds in the wild were breeding, and the first accurate census count--20 birds--was completed by photographing individual molting patterns (Snyder and Johnson, 1985; Snyder and Snyder, 1989). In 1981, a pair that had lost its first egg was observed to re-lay; this convinced authorities that a captive population could be successfully established by removing first eggs without harming the wild population. Permission to begin radiotelemetry studies was granted in late 1982. In spring of 1983, the first eggs were removed from every nesting pair, and four young were successfully hatched at the San Diego Zoo (Kuehler and Witman, 1988). Because condors only lay every 2 years if they are rearing a chick, two wild nestlings were also taken into captivity to ensure that the adults would breed again the next year (Snyder and Snyder, 1989). Field studies soon revealed that one of the principal causes of mortality was lead poisoning that resulted from the ingestion of lead bullet fragments in deer carcasses, and the sudden loss of six wild adults during the winter of 1984-1985 convinced many that all the wild birds should be brought into captivity. That proposal culminated in an expensive lawsuit against the USFWS by the National Audubon Society. The Court found in favor of USFWS. The last wild condor was brought into captivity in 1987, and the captive population at the San Diego Wild Animal Park and Los Angeles Zoo was thus raised to 27 birds. Captive condors began breeding in 1988, and the captive population has grown steadily to its current number of nearly 50 birds.

The California Condor Recovery Team has established three criteria that must be met before individual condors will be eligible for release to the wild: at least 90% of the genetic material of the founders, from which the released birds have descended, would still be represented in captivity; at least three birds must be physically and behaviorally ready for release at the same time; and at least three pairs must be successfully reproducing in captivity (Kiff, 1986, 1989; Wallace 1990). In anticipation of successful propagation of the California Condor, Andean Condors (*Vultur gryphus*) were used as research surrogates to develop methods for releasing captive-reared Condors to the wild. A total of 13 Andean Condors were released in southern California between 1988 and 1990 to test release and supplemental feeding procedures (Wallace, 1990, 1991), and the first captive-raised California Condors were released in the fall of 1991.

Magpie Robin (*Copsychus sechellarum*)

Once widespread and common on six islands in the Seychelles, the Magpie Robin by 1960 was restricted to Frigate Island, primarily as a result of the introduction of feral cats and rats (*Rattus* spp.) on the other islands (Gretton et al., 1991). Beginning in 1977, with about 41 birds remaining in the population, Jeffery Watson initiated a detailed field study under the auspices of the International Council for Bird Preservation and the World Wildlfe Fund to determine the factors responsible for the species decline. He found that predation on adults by feral cats was a major problem, and personnel of the New Zealand Wildlife Service conducted a successful eradication program. Later investigations by Jan and Mariette Komdeur indicated that the population was failing to increase because of several additional factors, including nest predation by skinks, snakes, and the introduced Indian Mynah (*Acridotheres tristis*), a shortage of suitable nest cavities, and a shortage of feeding sites resulting from the regeneration of secondary scrub forest. Consequently, a recovery plan outlining potential management activities was prepared and endorsed by the Seychelles government and Frigate's owner in 1990, and implementation began the next year. Management activities--including supplemental feeding, provisioning and protection of nest sites, and habitat improvement and forest restoration--appear to have stemmed the population's decline, but remaining population is very small (22 birds) and vulnerable to catastophes, such as fire, disease, and storms. Plans are now under way to establish a second population on another island as soon as the population increases to 30 birds. As with the Seychelles Brush Warbler (*Acrocephalus sechellensis*) (See Table 5.2), this action will be accomplished by translocating wild birds into suitable habitat (Gretton et al., 1991).

Lord Howe Island Woodhen (*Tricholimnas silvestris*)

This flightless rail is endemic to the subtropical Lord Howe Island in the southwestern Pacific Ocean. Although once numerous and widespread across this small subtropical island, the Woodhen by 1969 was reduced to about 20-25 birds on the heights of Mt. Gower as a result of the introduction of dogs, cats, rats, and goats. The remnant population of birds had survived on Mt. Gower in extremely poor habitat. The area was free of predators, but was marginal in all other respects (Fullagar, 1985). Field studies were initiated in 1971 to determine the status and ecology of the species, and a vigorous predator-control program was begun in 1976. Both programs continue. Beginning in 1980, a captive-breeding facility was established, and three wild pairs were trapped for breeding purposes. A total of 78 captive-raised and 13 wild-origin Woodhens were eventually released at various locations throughout the island, and by June 1984 the wild population exceeded 150 as a result of substantial reproduction (Lourie-Fraser, 1983; Miller and Mullette, 1985).

Takahe (*Notornis mantelli*)

Populations of this flightless gallinule, endemic to the alpine tussock grasslands of the Murchison Mountains on the South Island of New Zealand, declined precipitously as a result of predation by introduced mammals, competition with introduced deer for its principal food plant, and egg predation by Wekas (*Gallirallus australis greyi*) (Williams and Given, 1981). The bird was thought to be extinct until the rediscovery of a single population numbering about 200 birds in the alpine grasslands near Te Anau on the South Island, but this population declined to about 120 birds by 1983. Biologists determined that Takahes normally lay two eggs but raise only a single offspring, so in 1985 they began removing a number of eggs from the wild each year to establish a captive-breeding population and to hatch and rear young for release in other suitable areas. Between 1987 and 1989, 25 birds were released in the Glaisnock catchment of the Murchison Mountains, and additional populations were established on two predator-free islands. As a result of those manipulative procedures, the Takahe numbers increased to over 200 in 1990 (Hay, 1990; Reed and Merton, 1991).

Kakapo (*Strigops habroptilus*)

This large, flightless parrot, endemic to New Zealand, is unusual in that it is nocturnal, solitary, and polygynous with males forming leks for mating. Although at one time widespread on all of New Zealand's main islands, by 1976 it was believed to be effectively extinct, inasmuch as only a few males could be located (Merton, 1975). However, in 1977, a population of 100-200 birds was discovered on southern Stewart Island. Because predation by introduced mammals was a major impediment to recovery, biologists with the New Zealand Wildlife Service moved about 50 birds to three predator-free offshore islands beginning in 1982, and additional birds have since been moved between islands. When studies determined that the absence of reproduction in the translocated populations might be the result of inadequate nutrition, a supplementary feeding program was initiated in 1989 on Little Barrier Island. In 1990, male reproductive activities increased, and two females that were known to take the supplemental provisions (i.e., sweet potatoes, apples, brazil nuts, almonds, walnuts, and sunflower seeds) nested and laid eggs. In 1991, four of five females receiving supplemental food laid eggs, and two renested. Efforts to effect successful reproduction through supplemental feeding continue, and the populations are being closely monitored (Triggs et al., 1989; Moorehouse and Powlesland, 1991; Wise, 1991).

Chatham Island Black Robin (*Petroica traversi*)

Once widespread throughout the Chatham Islands of New Zealand, the Black Robin disappeared from all the larger islands after European colonization, until only a remnant population of 25 birds survived on Little Mangere Island. As a result of habitat degeneration in the 1970s, the population declined from 18 in 1972 to seven birds (two females and five males) in 1976, and the population was later transferred to 4 hectares of suitable habitat on Mangere Island. By 1979, only one female and four males remained, and an intensive effort to augment reproduction by removing eggs and stimulating renesting was begun. The eggs were initially cross-fostered to incubating Chatham Island Warblers (*Gerygone albofrontata*), but this effort was abandoned when it was determined that the chicks could be raised only to about 10 days of age. Eggs were later transferred 15 km by sea to South East Island and fostered to nesting pairs of Chatham Island Tomtits (*Petroica macrocephala*). To increase nest security and allow intensive manipulation, all Robin and Tomtit nests were slowly transferred to nest boxes. First and second clutches were generally removed from Robins for fostering, and third clutches were left with the Robins for incubation, hatching, and rearing. Owing to improper song-learning by male Robins raised to independence by Tomtits, fostering procedures were refined to return chicks to Robin nests at about 15 days of age, and supplemental feeding of both Robins and Tomtits was instituted to enable the parents to raise enlarged broods. Between 1980 and 1988, transfers between islands included over 40 Robin eggs, 10 nestlings, and 25 independent birds. A second population of Black Robins has been established on South East Island, and the total wild population has increased to over 100 birds (Flack, 1977; Reed and Merton, 1991).

Mauritius Kestrel (*Falco punctatus*)

Once common almost throughout the island of Mauritius, by the early 1970s the kestrel was restricted to native forest in the Black River Gorges region as a result of deforestation, and the known population consisted of only two breeding pairs. Like the 'Alala, this species existed for several decades at an effective population size of fewer than 20 pairs, and it was the subject of a poorly managed breeding program for nearly 10 years before success was achieved and significant reintroductions were begun (Jones et al., 1981, 1991; Cade and Jones, in press).

From 1974 to 1986, the wild population grew slowly and irregularly from two known pairs to about eight pairs (one to six productive pairs). The species essentially survived on its own during those years, despite attempts at captive breeding. Eight birds (five adults, and three nestlings) were removed from the wild for captive breeding between 1974 and 1978, but only a single young was produced in captivity. All nine birds died from one cause or another by 1980 (Jones, 1987). Although many conservationists lost hope for the kestrel, the program came under new direction when Carl Jones, of the Jersey Wildlife Preservation Trust, arrived on the island in late 1979. To minimize the impact on the wild population, he rebuilt the captive flock by removing first clutches for artificial incubation and rearing, thus allowing the wild pairs to renest and fledge young. The first captive-produced young were produced in 1984. With the help of Willard Heck, of the Peregrine Fund, more than 30 young had been produced in captivity by 1987, and more than 140 by 1992 (Cade and Jones, in press).

As the number of breeding pairs increased in the wild during the 1980s, further manipulations to enhance productivity were implemented. A total of 146 fledglings derived from 190 fertile eggs collected from the wild were hatched in the laboratory and later returned to the wild by fostering or hacking (Sherrod et al., 1981). In all, some 230 captive-produced and wild-origin young have been released since 1984. By the end of the 1991-1992 breeding season, the wild population totaled some 170 and included 30 known breeding pairs distributed in four areas of the island. Many of the birds that have been returned to the wild have survived and bred in extensively modified habitat (Cade and Jones, in press).

Conclusions

The recovery programs described here demonstrate a number of important points that are especially relevant to preservation efforts for the 'Alala:

- Wild populations of many threatened species have declined to extremely low numbers, but have responded positively to well-conceived and carefully implemented programs. The case studies cited here, as well as those for many other species, clearly indicate that, as long as a single male and female survive, it is never too late to try to save an endangered species. Most seemingly hopeless cases turn out to be much more hopeful when imaginative research and conservation programs are implemented (Derrickson and Snyder, 1992).

- The application of intensive in situ manipulative techniques can often lead to more effective and less-expensive means of augmenting wild populations than captive-breeding programs, although both approaches might be required initially until the number of wild breeders is increased significantly. Many recovery programs have substantially enhanced fecundity or survivorship of wild birds through the judicious application of predator control, supplemental feeding, and multiple clutching.

- The geographic distributions of several endangered species that had been reduced to single, relict populations have been expanded through deliberate releases into suitable habitat either within or outside their historical ranges. Some species, such as the Mauritius Kestrel, have even been induced to colonize successfully habitats that differ from their original ones. Not surprisingly, many programs have focused their efforts on reestablishing multiple populations (a "metapopulation" structure) (Simberloff, 1988; Murphy et al., 1990) to minimize the possibility of extinction caused by catastophic events.

- Suitable habitat is necessary for the re-establishment of wild, self-sustaining populations. Indeed, a recent analysis of vertebrate translocations (Griffith et al., 1989) determined that habitat quality is one of the most important predictors of success. Habitat preservation or restoration must have a high priority in every recovery program. This finding serves as a crucial reminder that endangered species represent signals, or "indicators" of systemic degradation in whole habitats or ecosystems. Habitat protection and restoration ultimately are the goals of any meaningful endangered-species recovery program.

In summary, on the basis of the successful restoration efforts that have been undertaken with a variety of avian taxa, the committee believes that there is a good chance that an effective program for the ‘Alala can still be designed and implemented. Options for programs that manage both wild and captive populations as one unit are discussed in Chapter 6.

6

OPTIONS FOR MANAGEMENT OF THE 'ALALA

Success in endangered-species preservation is possible only when the right combination of people and techniques are allowed to operate with adequate financial support and minimal institutional and political interference. The key ingredients for successful recovery programs are imaginative planning, effective administrative organization and commitment, funding continuity, creative and dedicated staffing, and effective communication and cooperation among interested and affected parties. Most of those programmatic ingredients are obvious, but the critical importance of coordination and cooperation needs to be emphasized. Recovery programs are never conducted in a vacuum, and they typically affect, directly or indirectly, diverse interest groups represented by government agencies, academic institutions, conservation organizations, and private landholders. Without the support and cooperation all of the parties, decisions will usually be made in the political rather than biological arena, and recovery actions will be thwarted or delayed.

As noted in Chapter 5, decisions about what to do to save critically endangered species involve complex considerations of biology, societal concerns, and government regulation. In selecting a course of action, it is often helpful to lay out all the basic options and to consider the positive and negative ramifications of each. Indeed, the techniques and processes of "risk analysis" or "decision analysis" are receiving increased attention from conservation biologists (Maguire, 1991; Starfield and Herr, 1991). Table 6.1 summarizes eight options for joint management of the two subpopulations of 'Alala and presents their main pros and cons. By comparing the advantages and disadvantages of these options, we might be able to clarify the tradeoffs among alternative courses of action and thereby reach a consensus (or at least maximize agreement) on how to save the 'Alala and its habitat.

Option 1. Passive Management (Protection) of Wild Population

This is essentially the only action that has been followed for the last decade to aid the wild crows. The relict population on the McCandless Ranch appears to have remained relatively stable during this period. Passive management could continue to maintain the status quo as long as natality balances mortality, or it could result in a slow increase in local numbers if natality exceeds mortality, as in the population model developed in Chapter 2. Passive management will *not* result in a rapid increase in the number of crows or in a major expansion of their geographic

Table 6.1 Potential management options for Hawaiian Crow

Advantages	Disadvantages
1. PASSIVE MANAGEMENT (PROTECTION) OF WILD POPULATION	
a. Prevents harmful influences and mortality related to human activity	a. Ignores conservation methods successful with other avian species
b. Provides compelling reason to maintain suitable habitat	b. Precludes veterinary support or monitoring of disease
c. Affords opportunity to obtain some biological information	c. Prevents collection of data necessary for managing wild and captive populations
	d. Does not protect small population from chance extinction
2. REMOVAL OF ALL BIRDS TO CAPTIVITY	
a. Might increase genetic diversity in captive population	a. Increases risk of losing species as a result of a local catastrophe
b. Increases size of captive population	b. Diminishes responsibility to protect the habitat
c. Protects last remaining birds from natural mortality in the wild	c. Eliminates habitat use and social traditions necessary for successful reintroduction
d. Increases potential for research studies of captive birds	d. Adult wild-caught birds are less likely to breed in captivity than in wild
e. Provides a source of birds for reintroduction	e. Might increase risk of transmitting diseases between wild and captive birds
	f. Eliminates acquisition of essential information on wild birds

Table 6.1 [continued]

Advantages	Disadvantages
3. TRANSLOCATE ALL WILD BIRDS TO ANOTHER LOCATION IN THE WILD	
a. Could allow research and management under more favorable sociopolitical conditions	a. Creates a potentially harmful effect of trapping and handling
b. Theoretically, birds could be moved to a more favorable environment	b. Birds might fail to acclimate to new situation and survive
	c. Provides no benefits unachievable in other ways
	d. Removes incentives to maintain existing habitat
	e. Factors responsible for the decline of species in areas of former distribution would not be identified or addressed
4. REMOVAL OF EGGS FROM WILD POPULATION FOR ARTIFICIAL INCUBATION	
a. Should increase hatchability and survivability of eggs and chicks and therefore increase reproductive output	a. Optimum techniques for artificial incubation are not yet established for the crow
b. Should increase the size of the annual cohort of young because Crows are capable of renesting and laying more than one clutch	b. Hatching and rearing chicks close to nesting birds will require new staffing and facilities
c. Allows determination of fertility and causes of embryo death	c. Some nesting pairs might not lay another clutch of eggs when first clutch is taken
d. Provides method to increase number and genetic diversity of captive birds	
e. Provides birds for enhancing or establishing wild or captive population	
f. Provides opportunities to conduct field research on released birds (e.g., with radiotelemetry)	

Table 6.1 [continued]

Advantages	Disadvantages
5. REMOVAL OF NESTLINGS FROM WILD POPULATION	
a. Circumvents uncertainties and facilities required for artificial incubation	a. Does not increase the gross reproductive output of the overall population
b. Increases the number and genetic diversity of captive population	b. Greatly decreases the possibility of renesting and producing more eggs
c. Increases number of birds available for release in unoccupied territory	
d. Might increase survival rate of young raised in the wild (fewer mouths to feed)	
6. FOSTERING AS A METHOD OF RELEASE	
a. Is the simplest and least expensive way to release captive-reared	a. Nestling mortality in the wild decreases effectiveness of fostering birds (e.g., caused by predation, disease, sibling competition)
b. Allows young to interact naturally with adult, parent crows	b. Opportunities with small number of wild pairs are few
7. HACKING AND OTHER "SOFT" METHODS OF RELEASE	
a. Allows release in vacant range where species is not already present (creates new populations)	e. Provides opportunity for establishing new behavior patterns and social organizations more adaptive to changed environments
b. Circumvents early-nestling mortality and in some cases first-year mortality, depending on when birds are released	a. Is labor-intensive and expensive
c. Allows release of large numbers of birds independently of wild birds	b. Naive birds held too long in captivity might lose capacity to adjust to wild conditions
d. Allows for adequate prerelease quarantine and screening for parasites and disease	c. Captive birds might not learn essential social traditions and survival skills needed in the wild

Table 6.1 [continued]

Advantages	Disadvantages
8. EXCHANGE OF CAPTIVE AND WILD CROWS	
a. Might increase genetic diversity of wild and captive populations	a. Increases the risk of disease transmission between populations
b. Allows captive and wild birds to be managed as one population	b. Captive birds might not survive and reproduce well in the wild
c. Could aid in formation of viable wild pairs	c. Wild adult birds are less likely to breed in captivity than in the wild
	d. Captive-raised birds might show serious behavior deficits (e.g., in foraging, predator recognition, and interactions with conspecifics)
	e. Does not increase the gross reproductive output of the overall population
	f. Requires close synchrony in breeding of wild and captive pairs

range. The expansion of their range, through the establishment of additional, separate populations ("metapopulation") (Levins, 1970; Soulé, 1987; Simberloff, 1988), is one of the main strategies needed for long-term survival of the species. Moreover, for the foreseeable future, an unmanaged crow population will remain so small that it will continue to be subject to a high probability of extinction from chance events related to genetics, demography, or local environmental catastrophes.

It is true that passive management provides a compelling reason to maintain the crows' existing habitat and affords some opportunity to obtain needed biological information unobtrusively (e.g., determination of incubation time, nestling period, social structure, and seasonal movements and their biological functions), but those benefits are not precluded by some other options that emphasize active management. The possible prevention, by passive management, of harmful disturbances or deaths resulting from human actions must be weighed against the benefits of manipulation. Although there is some evidence that research-related disturbances during past breeding seasons caused pairs to abandon their nests, such disturbances can be minimized by the use of proper research techniques and by progressive conditioning of birds (cf. Grier and Fyfe, 1987). The benefits to species survival of active management far outweigh the occasional losses of individual birds associated with accidents that result from research or management.

When its advantages and disadvantages are compared, Option 1, passive management, evinces no compelling benefits as the exclusive mode for conserving the 'Alala. It is time to replace passive management with intelligently considered, and carefully executed, active management.

Option 2. Removal of All Birds to Captivity

Translocation of all individuals of a species from the wild to captivity is an extreme measure that has been carried out on only a few endangered species--the California Condor (Snyder and Snyder, 1989), the black-footed ferret (*Mustela nigripes*) (Thorne and Williams, 1988; Clark, 1990), the Guam Rail (*Rallus owstoni*), and the Micronesian Kingfisher (*Halcyon c. cinnamomina*) (Witteman et al., 1990). In those cases, the surviving populations were so small (10-30 birds remaining in the wild) and the adult mortality so abnormally high that extinction within a few years could be confidently predicted. It should be noted that the captive populations of the condor, ferret and rail have thrived with high reproductive success; and the first exploratory releases of captive-produced animals have been initiated in the last 2 years. It remains problematical, however, whether any of these species can be re-established in the wild in the absence of social organization or a tradition of habitat use into which animals released from captivity could fit, although Andean Condors might function as guides for the released California birds. For some species with relatively simple social systems and largely innate (instinctive) responses to environmental factors--such as food, nest sites, and predators--re-establishment of naive animals into vacant habitat is relatively easy; the Peregrine Falcon (*Falco peregrinus*) is an example. For other species, with highly developed social organizations and a dependence on learning to cope adaptively with environmental factors, reintroduction is much more complicated and has a lower probability of success (Lyles and May, 1987; Snyder and Wallace, 1987; Scott et al., 1988; Snyder and Johnson, 1988; Griffith et al., 1989, 1990; Derrickson and Snyder, 1992).

Removal of all 'Alala to captivity has been seriously discussed. Certainly the number of surviving crows is critically low, and high adult mortality has been implicated in this report as a more important factor than poor reproduction in the decline of previous decades (Chapter 2). It is hard, therefore, to argue against any of the advantages listed in Table 7.1. However, the amount of genetic diversity that would be introduced into the captive population by Option 2 would depend on how distantly related the wild birds were to the captive birds, and those relationships are unknown. (It is likely that all surviving 'Alalas are inbred to some degree--see Chapter 3.) Protecting all birds from high mortality in the wild is the only benefit that could not be accomplished by other options, but adult survivorship in the single remaining wild population appears to be high (see Chapter 2).

Some of the disadvantages of Option 2 could be partly mitigated. Protection from catastrophic events in captivity could be increased by housing the crows in multiple facilities. Legally binding agreements could be reached about protecting the vacated habitat for future reintroduction. Improved veterinary services could greatly reduce the risks of diseases in captivity. The option would, however, absolutely preclude acquisition of further information about crows in the wild and would destroy existing social traditions and learned habitat uses specifically related to the McCandless Ranch, the only environment that is certainly still suitable for crows.

For several reasons, Option 2 should be viewed as a *last resort*, to be undertaken only if it becomes certain that adult mortality exceeds recruitment and that the number of breeders is clearly in decline. First, the available data (Chapter 2) indicate that the relict population of crows on the McCandless Ranch has been relatively stable in number for the last decade and might even be increasing slightly and producing a surplus of nonbreeders for potential dispersal or new territory establishment. Second, there is a great need to obtain information on the biology of crows in the wild that is critical for management. Third, like other corvids, the 'Alala appears to have a relatively complicated social system and is probably a species that depends on transferring adaptive information and habits from parents to offspring. Although leaving the remnant population in the wild constitutes a risk whose magnitude depends on the uncertainties inherent in the demographics of small populations and on unpredictable environmental catastrophes, it appears that the disadvantages of Option 2 outweigh the advantages, especially because most of the advantages can be achieved through other options that depend on retaining a viable population of wild crows.

Option 3. Translocation of All Wild Birds to Another Location in the Wild

This extreme measure has been used successfully to rescue the Chatham Island Black Robin from probable extinction in New Zealand (Reed and Merton, 1991). It has also been proposed for the 'Alala, primarily to circumvent sociopolitical problems that have hampered state and federal efforts to manage the wild crows, most or all of which now live on private land. From a regulatory and institutional standpoint, there could be advantages to having the crows on public land, rather than private land. If the new location were biologically more suitable than the current range, that would be an added incentive for such a translocation. However, there is no certainty that the crows would adapt to a new location; more important, translocation of all crows provides no biological or managerial benefits that could not be achieved in other ways. Option 3 is not recommended, although once numbers are increasing, the translocation of some birds to a new areas could be a reasonable future option, as a method of establishing additional populations. Cooperation of private landowners would need to be an integral part of the recovery plan.

Option 4. Removal of Eggs from Wild Population for Artificial Incubation

Egg removal ("egg-pulling," "double clutching," or "multiple clutching") is an increasingly common method for augmenting the natural reproductive output of wild birds (Carpenter and Derrickson, 1981; Derrickson and Carpenter, 1987; Cade et al., 1988; Kuehler, 1989; Snyder and Snyder, 1989; Wallace, 1990; Jones et al., 1991; Reed and Merton, 1991; Derrickson and Snyder, 1992). It relies on the capability of most female birds to renest and lay a second or third set of eggs after removal of the previous clutch or to continue laying eggs beyond the normal clutch size if eggs are removed in sequence as laid. Because eggs have a much higher rate of hatch in laboratory incubators if they have received several days of natural incubation, taking full clutches after 5-7 days is the best procedure.

The great advantage of Option 4 is that it provides a way to take substantial numbers of eggs from the wild for hatching and rearing of young without depriving the wild birds of all their reproductive output. There are two slight disadvantages for the wild birds: a higher percentage of late-laid eggs than early eggs are likely to be infertile or otherwise unhatchable, and late-hatching young might not survive as well as earlier ones, because of seasonal changes in food supply (although this can be alleviated by supplemental feeding) or because of other changing factors, such as weather (e.g., higher temperatures and more rain). The important point, however, is that the combination of natural and artificial hatching and rearing yields many more young than could be produced by the unaided wild population. The young reared in captivity can be used in several ways: they can be released back into the wild, they can be retained in captivity to augment the captive breeding stock, and they can be used to learn more about the biology of the species.

Like other corvids, the ‘Alala has been documented to renest after loss of a first clutch (for reports on ‘Alala see Temple and Jenkins, 1981; for recent reports on other species see Butler et al., 1984; Stiehl, 1985; Kilham, 1986a; Buitron, 1988; Goodburn, 1991). Efficiency and success would be increased by placing the incubation and rearing facilities near the wild population. Even collected eggs that are infertile or that die during incubation can provide important information on the effects of inbreeding on reproduction and on the role of nutritional influences in hatchability and embryonic development.

Although it is true that optimum conditions for artificial incubation have not yet been worked out for the ‘Alala, this problem can be partly offset by making sure that the wild eggs receive some natural incubation before they are removed. Modifications of the current method of incubation at Olinda suggested in Chapter 4 should also improve the hatchability of both wild- and captive-produced eggs. It is unlikely, on the basis of what is known about other corvids and birds generally, that failure to renest will be more than an infrequent occurrence or a rare peculiarity of particular females.

Option 4 should receive a high priority, because it provides a way to achieve most of the management objectives through the simultaneous buildup of a captive-breeding colony or colonies, and the manipulative establishment of captive birds in the wild, without necessitating the removal of any juveniles or adults from the wild population or otherwise seriously interfering with the social organization and productivity of the wild crows. Removal of eggs with the later option of reintroduction obviously constitutes a long-term strategy, and a working recovery team should evaluate the results annually and make recommendations about what to do with the crows that are produced in captivity on the basis of careful consideration of the demographic and genetic requirements of both the wild and captive populations.

Option 5. Removal of Nestlings from Wild Population

This option provides essentially the same products as removal of wild eggs, but without the labor and cost of maintaining facilities for artificial incubation of eggs. Its chief drawback is that it does not increase the gross reproductive output (number of fertile eggs) of the combined populations and can exert little influence on net reproductive output by increasing the survival rate of nestlings. It should be considered in addition to egg removal, especially for wild pairs that have a history of poor nestling survival--e.g., a pair that usually hatches two or three young, but consistently raises zero or one to fledgling age. Obviously, the relevant data on reproductive history have to be in hand in order to use Option 5 intelligently.

Option 6. Fostering as a Method of Release

Fostering young into the nests of wild "parents" is the simplest, least expensive, and most natural way to reintroduce captive-reared birds into the wild and it has been used with great success for several altricial (helpless at hatching) and semialtricial species, especially raptors (Cade et al., 1988; Garcelon and Roemer, 1988). At least initially, it would have little applicability to the 'Alala because the breeding pairs are few (currently estimated to be four or five) and because pairs of crows appear able to fledge only one or two young. Supplemental feeding might improve that number to the point where three or four young could be reared per nest, but this possibility should be tested first on unaugmented, natural broods of two or three. If wild pairs or groups of crows can be induced to adopt fledglings instead of nestlings, then additional opportunities for efficient augmentation will be possible.

Cross-fostering into the nest of a different, surrogate species is a variant of fostering that can be used when a suitable surrogate species is available in a suitable environment. However, there is not now a suitable surrogate--native or exotic--for the 'Alala. If the 'Alala requires the transfer of complicated social traditions or survival techniques from parents to offspring, or if social and sexual imprinting are critical to the development of normal functions, then cross-fostering would be precluded as a management technique.

Option 7. Hacking and Other "Soft" Methods of Release

It is highly unlikely that a "hard release" (turning birds loose in a new environment) would work for captive-reared crows. Some form of slow, controlled release over a period of days or weeks (hacking) will probably be needed to allow captive-reared crows time to acclimate to wild conditions and to learn the necessary survival techniques--which foods to eat, how to avoid predators, where to roost, and so on. These "soft" techniques range from the traditional "hacking" used for raptors (Sherrod et al., 1981) to more elaborate systems involving the holding of birds for extended periods in outdoor flight cages in the habitat where they are to be released (Broo, 1978; Bloesch, 1980; Zwank and Derrickson, 1981; von Frankenburg et al., 1984; Blackwell, 1990). Depending on the species and the particular demands of the environment, the birds can be released at any age from fledgling to sexually mature adult. They might be released as individuals, as "sibling" groups, as integrated flocks, or as paired adults.

In the case of the ʻAlala, it is likely that results will be best with the release of social groups--perhaps as juvenile flocks with one or more wild or captive-produced adults as flock "guides." They might be released as paired nonbreeders, or even as breeders that are first allowed to nest in a holding enclosure before being set free. Several different approaches will need to be tried to discover which works best.

The main advantage of soft procedures is that they provide a way to establish released birds in vacated habitat where no wild population exists. Thus, new populations can be created, the range of the species becomes expanded, and the species becomes reorganized into a metapopulation (linked through dispersal) and with a higher probability of long-term survival. Soft releases can sometimes also avoid the high mortality associated with fostering eggs or young birds. Moreover, they provide opportunities to acclimate and condition birds to survive and function adaptively in degraded or partially exotic habitats that unconditioned, wild birds would find difficult or impossible to live in (Cade, 1986a; Cade and Jones, in press).

The main drawback of Option 7 is that soft releases require a cadre of dedicated and informed workers and much money over a long period. In addition, naive, captive-reared crows can present special problems related to socialization and learning, as noted in Thick-Billed Parrots (*Rhynchopsitta pachyrhyncha*) (Snyder and Wallace, 1987), Hispaniolan Parrots (*Amazona vittata*) (Wiley et al., 1992), and Mississippi Sandhill Cranes (Zwank et al., 1988). Captivity can also result in genetic, anatomical, or behavioral changes that can compromise survival in the wild (Derrickson and Snyder, 1992). Even so, some variation of Option 7 will be the only likely way to expand the species into now-unoccupied range and to increase substantially (say, by a factor of 10) the number of crows in the wild. Once field studies have clarified the social organization of the ʻAlala and the role that learning plays in the acquisition of social behavior and survival skills, it should be possible to design a soft-release procedure that works effectively.

Option 8. Exchange of Captive and Wild Crows

The main advantage of Option 8 is that it allows the small wild and captive populations to be managed as one population, so that gene flow can be maintained between the two subpopulations while increase in numbers is under way (recovery). A demographic advantage accrues to the wild population primarily through the exchange of wild eggs or nestlings (possibly some yearling and adult birds) for later captive-produced young, yearling, or adult birds that can be released (Options 4, 5, 6, and 7). Another demographic advantage would be to fill in existing "gaps" or vacancies in the wild population, as when a mate lost from a pair is not replaced by a wild bird. A case in point is the single female that might still survive in Hualalai. If a suitable captive male were available, he could be released in the new home range and be carefully monitored. Alternatively, she might be trapped and brought into captivity. In either situation, her genes could make an important contribution to the genetic diversity of the overall population, because she is likely to be more distantly related to the crows on the McCandless Ranch than any of them are to each other.

The main disadvantage to Option 8 is the potential for disease transmission. The adoption of strict quarantine and veterinary screening procedures should reduce that risk to a minimum, and such a precaution tips the balance in favor of this option.

Conclusions

It appears that a combination of several options would yield the highest probability of a successful recovery of the 'Alala. In general, the wild and captive populations should be managed as one population, with human intervention to direct gene flow and the exchange of birds between the two subpopulations on the basis of the best available information on genetics and pedigrees and on demographics of the two subpopulations. Unless an unexpected emergency arises, the wild population on the McCandless Ranch should be left essentially intact, i.e. no removal of adult or postnestling birds until numbers have increased substantially--at least a doubling of the effective population size. Manipulation of the reproductive output of the wild crows should begin in the 1993 breeding season with the removal of first sets of eggs from all nests that can be located. If possible, the eggs should be incubated and the young should be hatched at a facility near the wild population. Some of the young hatched and reared in captivity should be retained as captive-breeding stock; others should be experimentally released according to one or more of the methods outlined above. The released birds should be color-banded and radio-tagged so that the maximum amount of biological information about them can be obtained.

As the wild population increases over the next few years, additional manipulative procedures can be introduced--supplemental feeding of wild pairs; translocation of nestlings, subadults, and nonbreeding adults to new territories; development of optimum "soft-release" methods for establishing crows in a vacated range; perfection of methods to prevent disease in the wild and captive birds; and conditioning of crows to acclimate and adjust to human-modified habitats.

7

FINDINGS AND RECOMMENDATIONS

Chapters 2 through 4 describe what is known about the 'Alala, and Chapter 5 describes relevant precedents for active management of other endangered species. Chapter 6 describes in detail options for the joint management of the wild and captive populations of 'Alala. Having examined the available data and studied the experiences of others working with endangered avifauna, we now offer a summary of what has occurred to date and some recommendations for action. It is imperative to build on what is already known about the 'Alala and other species if we are to make informed decisions about recovery and management actions.

The Wild Population

Findings

The 'Alala--an omnivorous, but primarily frugivorous, forest-inhabiting corvid that differs in those respects from the widespread and familiar crows of the larger continents--is near extinction. Its decline is part of a larger phenomenon: reduction and extinction of forest birds, especially frugivores, that have been associated with human colonization throughout Polynesia.

As of January 1992, the 'Alala continues to exist on the McCandless Ranch, which are present in two and possibly three territories. A count in March and April 1992 located 11 birds. A few additional 'Alala might persist elsewhere on the Kona Coast north of the McCandless Ranch, and perhaps within the Ka'u District, but no other breeding population is likely to exist.

In the last 14 years, death rates of adult 'Alala in the wild were inordinately high, except at the McCandless Ranch, where numbers of 'Alala observed appear not to have changed since 1976. We do not know why adult survivorship has been low or why it has been higher at McCandless than at other sites, but this fact alone can account for the disappearance of 'Alala from two areas (Hualalai and Honaunau).

Limited banding data and census efforts in the Kona District during the 1970s and 1980s indicate that during this period of precipitous decline, clutch size for the 'Alala is at the low end of what has been found to be typical of that of other *Corvus* species; nesting pairs of 'Alala continued to produce fledglings at the McCandless Ranch at rates slightly lower than those of other corvids, and even lower elsewhere; and juvenile survival (up to 1 year) in the Kona District was comparable to that of other corvid species that are not endangered.

FINDINGS AND RECOMMENDATIONS

Although extensive field studies have been conducted, many aspects of the behavior and biology of the 'Alala are still poorly understood, and no single cause for the decline in the wild 'Alala population can be identified. However, previous studies implicated habitat and food, predators, and diseases and parasites. When populations are as small as this one, demographic accidents and random environmental disturbances are likely to cause extinction.

Clearly, a dominant cause of the decline has been alteration of the 'Alala habitat throughout its historical range as a result of grazing by ungulates, logging, and agriculture in the mid-elevation belt (1,000-1,800 m).

The magnitude of mortality caused directly by collecting and shooting is difficult to know, but it has probably been substantial, even in recent decades. Furthermore, introduced roof rats prey on eggs and nestlings, and introduced mongooses prey on fledglings on the ground.

Two introduced diseases that are widespread among native birds and affect the 'Alala are avian malaria and avian pox.

The number of crows that would constitute a minimum viable population for long-term survival of the species cannot be estimated reliably at this time. But it can be said that the species will be in danger of extinction for the foreseeable future.

The committee is unanimous in believing that without an active management program the 'Alala population on the McCandless Ranch is likely to become extinct in 1-2 decades from chance, genetic, or demographic events.

Maintaining a Wild Population

Recommendation 1: ***In light of what is known about the 'Alala, a viable wild population should be established and maintained.***

The current size of the population in the wild must be increased for demographic and genetic security. Considering how few crows remain, it is advisable to manage the wild and captive populations as a single unit. Such joint management will require that the identity of all existing birds be known. The possibility that additional birds survive on public and private lands in Hualalai and the Ka'u District needs to be investigated by the U.S. Fish and Wildlife Service and the state of Hawai'i, because additional birds would be potentially crucial to species recovery. They could provide critical, traditional experience for newly released, naive 'Alala being reintroduced into formerly occupied territories.

However, contingency plans for the wild population must be developed in the event that the strategy proposed above fails. While the wild population must be maintained if at all possible, there might come a time when it would be necessary to remove all birds from the wild and put them in captivity (Option 2). This option should be considered only under the following circumstances: 1) when the wild population is known to be less than two breeding pairs for two consecutive years, and 2) state-of-the-art breeding facilities exist on the Hawaiian islands and the captive population is reproducing well.

Recovery Team

Recommendation 2: ***A new recovery team or advisory working group for the 'Alala should be established.***

The federal Alala Recovery Plan (Burr et al., 1982) and the state Alala Restoration Plan (Burr, 1984) are admirable documents that contain concise summaries of the history and status of the species and sound recommendations for recovery. Each places priority on the protection and restoration of native habitat, the study of disease, and predator control, and each recommends an integrated management of the captive and wild populations. The federal plan designates critical habitat that needs protection and the state plan sets priorities for incorporating that land into a conservation scheme. In the decade since the federal plan was issued, however, it has not been implemented, and the 'Alala has declined further.

The committee recommends establishing a new recovery team or advisory working group for the 'Alala that includes state and federal biologists; other professional biologists who are experts in ecology, captive propagation, reintroduction, population biology of birds, and population genetics; an avian veterinary pathologist; an aviculturist; a corvid specialist; and a representative of the private sector, preferably a private land-owner or land manager. This combination of experts will provide the knowledge and skills necessary for joint management of the wild and captive populations. The recovery team must monitor the progress of the 'Alala's recovery and identify research priorities.

Land Management

Recommendation 3: ***For the 'Alala, as for other endangered species in Hawai'i and elsewhere, habitat restoration and maintenance are of paramount importance for species recovery and sustainability. Efforts to control and counteract the impacts of exotic animal and plant species must be given high priority.***

In other habitats along the Kona coast where the 'Alala has already been extirpated, numerous endemic bird populations are declining. Unless the causative factors are identified and corrected, preservation efforts for the 'Alala will be seriously compromised.

FINDINGS AND RECOMMENDATIONS

Habitat preservation

Recommendation 4: ***The committee strongly urges the state of Hawai'i and the U.S. Fish and Wildlife Service to establish one or more new preserves along the Kona coast.***

Acquisition, restoration, and proper management of a significant, dedicated ecosystem preserve on the Kona slope is important for the continued existence of 'Alala in the wild. As discussed in Chapter 2, little pristine forest habitat exists anywhere in the Kona District. The forested slopes of the Kona District constitute one of the most important centers of endemism in the entire Hawaiian Islands chain. Despite its enormous biological significance, the Kona District lacks a habitat preserve that is dedicated to and managed for the native habitat. The absence of a dedicated forest preserve in Kona is especially problematic if the 'Alala or any other endangered species is to be reintroduced. One cannot justify the release of captive-reared birds into areas where the native population has been extirpated and where there is no plan for substantial improvement in the long-term management of the native ecosystem. At best, it would effectively doom any reintroduction program to producing an expensive repetition of the original extirpation events. At worst, it would cause the program to fail to re-establish any reproducing population.

For purposes of 'Alala recovery, the recommended preserve should be as close as possible to the McCandless Ranch and should contain areas of forest both at the nesting elevations (1,200-1,800 m) and in the feeding zones (especially down to 800 m in moist 'ohi'a forest).

Recommendation 5: ***The committee strongly recommends that state and federal agencies and private land-owners develop a program of cooperative agreements, easements, and modes of direct compensation to promote a system of exclosures of various sizes and locations. It also recommends that such a program include other creative manipulations of the habitat to protect habitat patches on ranches from the impacts of cattle and other herbivores, so as to restore and maintain the essential features of crow habitat in sufficient quantity and quality for viable populations of crows and other native species of plants and animals to be secure for the indefinite future.***

During the 'Alala's recovery period, the cooperation of ranches in the vicinity of the remnant wild population will be vital. Research on the specific behavioral use of these properties by 'Alala will provide much needed insight into how cattle and 'Alala can coexist elsewhere along the Kona slopes. If the state or federal government can acquire a large tract of land for a dedicated Kona coast ecosystem, as recommended above, then it could function, in concert with a system of exclosures on surrounding ranch lands, as a reservoir from which species could disperse into peripheral habitats.

Cattle ranching

Recommendation 6: ***The committee encourages the economical management of cattle ranches in ways that provide critical habitat for native flora and fauna.***

Ranches that incorporate a holistic perspective abound throughout North America and are increasing there and in other parts of the world, e.g., Africa, South America, and Australia. Because ranches are large, and therefore can contain larger stands of native or near-native habitat, they make vital contributions to the long-term protection of landscapes and ecosystems, as long as they are managed in a benign, self-sustaining way with a multiple-use approach, and are not exploited for maximum short-term financial gain. Recovery and long-term survival of 'Alala through extensive portions of its native range depend on the good will and interest of the cattle-ranching community, which can be obtained only if ranchers understand how the economic goals of food production and healthy natural forest management can be met simultaneously.

Ranchers often know and understand the natural history of their properties extremely well, in part because they spend considerable time attending to the annual cycle of their landscape and animals. Property-owners involved with ranching operations usually are justifiably proud and fond of the natural attributes of their land. These persons need to be respected, listened to, and encouraged to participate fully in the process of understanding and protecting endangered species, such as 'Alala, and their habitats and ecosystems.

Four features of habitats, as set forth below, are crucial to the 'Alala. All four can be maintained in good quantity on cattle ranches. If property-managers control or prevent clearing and grazing in critical areas of sufficient size and distribution, 'Alala territories ought to be able to persist indefinitely in cattle-ranch settings on the Kona slopes.

- *Tall trees and forest patches*. 'Alala was always a forest-inhabiting crow with the habits of roosting, resting, displaying, and nesting in tall trees. 'Alala *will* live in forested areas that are interspersed with openings, pastures, occasional clear-cuts, and selective tree removals. 'Alala *will not* live in areas that have lost their essential forest stature. Mature 'ohi'a and emergent koa must exist in dense stands covering tens to hundreds of hectares. Numerous smaller wood lots widely dispersed and connected by corridors might be better than one or a few extensive stands.

- *Native fruit*. 'Alala consume a wide variety of fruits all year round and even feed them extensively to dependent young. Especially important traditional fruits include: 'ie'ie, mamaki, 'olapa, 'oha, ho'awa, and pilo. They and many other native fruits grow on shrubs, small trees, and vines. Quantities of those plants must be protected from clearing and grazing to provide feeding habitat for 'Alala. Moreover, the plants produce fruit for only part of the year, so 'Alala requires adequate supplies of diverse native food plants to prevent critical nutrient shortages during any season. In some small key areas (e.g., areas with certain fruiting shrubs), creative horticulture could be employed to re-establish populations of these plants in areas occupied by cattle and other ungulate herbivores by growing them as epiphytes in artificial planters suspended in trees above the heads of the livestock.

- *Understory vegetation.* Although often seen in the forest canopy, ‘Alala regularly forage within several meters of the ground (Sakai et al., 1986), and fledgling ‘Alala depend on understory and ground cover to shelter them from predators for several weeks following fledging. Native understory can be maintained only by excluding cattle and other grazing or rooting mammals. Maintenance of large examples of such terrain interspersed throughout a ranch property is especially critical for ‘Alala nesting.

- *Habitat corridors.* ‘Alala prefer forest and forest-edge habitats and will disperse more readily when provided unbroken habitat corridors. In a ranch setting, these can be maintained efficiently by excluding cattle from networks of forest corridors along fencerows, road margins, and natural forest-edge ecotones. It is especially vital that corridors of native ‘ohi‘a-koa forest be allowed to regenerate near the upper elevation of this habitat. Owners of neighboring properties could further enhance ‘Alala habitat by coordinating habitat corridors to connect with one another along the north-south axis of the Kona slopes, especially in the elevational zone between 1,000 and 1,800 m.

Predator control

Recommendation 7: ***The habitat of the wild population should be managed aggressively with respect to predators, including mongooses, rats, and cats.***

The current practice of trapping cats, mongooses, and rats on the McCandless Ranch should be continued and expanded. Because ‘Alala will frequently nest in the same locations from year to year, nest trees should be protected by electrical predator guards or aluminum flashing during the breeding season. Solar-powered electric guards have been used successfully on Guam to protect Marianas Crow nests from predation by brown tree snakes (*Boiga irregularis*). In addition, this year fledgling Marianas Crows were successfully protected during the vulnerable postfledging period by placing them in cages each night until they were capable of sustained flight (R. E. Beck, pers. comm., 1992). This technique should be tried with ‘Alala fledglings to reduce the risk of predation.

In addition to trapping, other methods, including the use of diphacinone-laced baits, should be considered when the recovery team feels that data are adequate to evaluate their safety.

Management of the Wild Population

Recommendation 8: ***Techniques must be developed for managing the wild population of 'Alala without risking injury to it.***

The goal of joint management of the captive and wild populations is to increase the size (density and distribution) of the population rapidly. The following short-term actions that can be instituted immediately are recommended:

- Experiment with effectiveness of food supplementation to increase nestling and fledgling survival. Determine important, safe, and practical kinds of food supplements through additional research.
- Promote survival of wild nestlings and fledglings with predator-control enclosures around nests aimed at mongooses, cats, and rats.
- Locate and promote the growth and preservation of trees that are optimal nest sites.
- Set up observation towers of the scaffolding type on the McCandless Ranch, e.g., four towers 1 mile apart and equipped with a high-powered, high-quality spotting scope and walkie-talkies to facilitate the documentation of habitat use, nest sites, behavior, and banding of young.
- Monitor and collect Kalij Pheasants and wild turkeys for disease analyses.
- Establish a public-education program for the 'Alala and other endangered forest species of Hawai'i.

The following long-term actions that require planning, cooperation, and coordination of several parties are recommended:

- Improve the 'Alala's habitat through forest management and other practices that promote important fruit-bearing plants and substrates for invertebrates.
- Institute a long-term study of avian malaria at appropriate elevations on the Kona slopes.
- Investigate the possibility of vaccinating nestlings and fledglings against avian pox in order to promote increased survival. Investigate the feasibility of treating young 'Alala for malarial infections.
- Augment reproductive effort by using young produced in captivity for reintroduction.
- Design a cage-within-a-cage aviary, with a rat-proof floor, for release of the fledglings on the McCandless Ranch.
- Release young hatched from first clutches back into the wild immediately. Use "soft release" methods to induce young crows to use habitat adjacent to existing 'Alala territories. They might even assimilate into newly forming groups on existing territories. Reintroduce crows in groups of two or three, not singly. Favor the McCandless Ranch region for releases until the numbers have grown substantially.

Recommendation 9: ***Beginning in the 1993 breeding season, remove first clutches of eggs from the nests 5-7 days into the incubation period and carry them by hand to a hand-rearing facility on the McCandless Ranch. Allow the 'Alala pairs to incubate and rear their own second clutch without interference. Locate the holding facility near the adult population where vocalizations of wild adult birds can be heard. Observe whether adult birds respond to the hand-reared fledglings.***

Additional Research

Recommendation 10: ***Additional field research is vital for the survival and recovery of the 'Alala. Increasing their numbers in the wild requires knowing what they need to live, remain healthy, reproduce, and recruit in the existing, highly modified environment.***

Some of the most basic biological facts are still unknown, especially regarding habitat needs, annual and seasonal food requirements, social behavior, and demographic characteristics. Most urgently needed are data on habitat requirements and the variation in food resources that are available and required through the seasons. Such data will provide specific direction for management of the property that supports the final wild population, of adjacent properties as 'Alala numbers increase, and of additional forest preserves after reintroduction or recolonization.

The following pages summarize the kinds of data and observations that are most needed to help guide management of habitat as 'Alala numbers increase. The data can be gathered noninvasively, with minimal disturbance of the wild birds.

Numbers

Surveys of all recent 'Alala ranges should be conducted every year, including broadcasted voice recordings during the winter and very early spring (i.e., during dispersal and early territory establishment, when 'Alala are most vocal and responsive). Effort should be invested at appropriate elevations (1,000-1,800 m) on adjacent properties--i.e., Honaunau Bishop Estate Lands and Yee Hop Ranch--to determine whether additional pairs, resident solitary individuals, or occasional dispersers exist in the vicinity of the McCandless Ranch population. Surveys should also be conducted in the Ka'u District.

Habitat

Actual and potential resources for 'Alala need to be studied throughout their recent range, not just where they live now. Replicate study plots should be established at the McCandless Ranch and on the best areas immediately adjacent to the ranch, where 'Alala might first be able to recolonize. In addition, at least three recently occupied sites remote from the McCandless Ranch should be similarly studied for comparison and to guide future management. Exclosures should be established as soon as possible for long-term studies on vegetational changes in the absence of impacts by grazing herbivores. Further experiments should be carries out to develop ways to re-establish native plants on ranges continuous to those occupies by livestock. Some areas should be totally enclosed using pig-control fences as have been used in Hawai'i Volcanoes National Park. Sites anticipated for future reintroductions should be of highest priority for study.

Quantitative analysis should be conducted of the differences between existing 'ohi'a-koa forests inside and outside the present range of 'Alala and on the McCandless Ranch inside and outside the home ranges of the two or three remaining 'Alala groups. Special attention should be placed on measuring the composition and abundance of plant species in the forest understory; forest structure, especially age of trees and dispersion of openings within the forest; ages and patch sizes of the lava flows underlying the forest and their effects on plant species composition.

Seasonal fruiting phenology and overall availability of native fleshy fruits (especially Lobeliaceae, Rubiaceae, and *Freycinetia arborea*) must be examined. If any resource appears to be disproportionately favored by 'Alala (e.g., some particular species of understory fruit), the limiting factors controlling dispersal and recruitment of this resource should be determined, especially the effect of grazing or rooting by cattle and pigs.

Characteristics of nest sites and their vicinity should be identified, especially quantitative descriptions of the structure of vegetation, plant composition, and extent of native understory habitat within 250 m of the nest trees. Those measurements can be made after the end of the nesting season to avoid possible disturbance of any active nest.

Foraging behavior

The proportions of time that the 'Alala spend in foraging and roosting in different habitat types (ground, understory, vines, and canopy,) compared with the availability of these habitats within the home range of family groups, should be measured throughout the year. Special attention must be paid to the forest structure and composition where 'Alala choose to spend their time, for example, whether or not open, partially logged 'ohi'a-koa forest is used as much as closed-canopy forest patches in proportion to its occurrence on the McCandless Ranch.

Food preferences of foraging 'Alala, compared with the availability of these resources, should be measured throughout the year. Of particular interest would be which fruits, if any, appear to be disproportionately favored.

The behavior of 'Alala during movements away from the permanent territory, especially during fledgling-feeding and during midwinter, should be monitored to determine what they are searching for when they leave the permanent territory and how important downslope food resources are during different times of the year.

FINDINGS AND RECOMMENDATIONS

Consideration should be given to supplementing the food supply on active territories in an effort to increase reproduction. Although initial attempts did not prove very successful (Banko and Banko, 1980; Giffin, 1983), additional experimentation is warranted. Food supplementation in a variety of species, including corvids (Yom-Tov, 1974; Hochachka and Boag, 1987; Hochachka, 1988; Dhindsa and Boag, 1990; Richner, 1992), has been shown to have a number of effects, including an advancement of laying, increased clutch size, and increased nestling survival and fledgling success.

Physiology and disease

The incidence and role of disease as a limiting factor among wild ʻAlala are poorly known. Disease could be a dominant factor in their decline, or it could be essentially irrelevant. The issue is vital to the long-term ecology of vertebrates on the island of Hawaiʻi. ʻAlala numbers are now too low for active study of the birds themselves in this respect. Inference through the study of other bird species is now required, and it is essential. The following are some questions that need further examination:

- What are the infection rates of avian malaria and avian pox among introduced and native forest birds at 1,500 m on the Kona slope, both on and off the McCandless Ranch? Does any aspect of disease incidence or vector ecology appear different on the McCandless Ranch?
- What are the primary and secondary breeding areas for disease vectors above 1,000 m in ʻohiʻa-koa forests, especially on and near the McCandless Ranch?
- Are introduced birds, such as the Kalij Pheasant and wild turkey serving as disease reservoirs for malaria, pox, or other diseases?
- What are the nutrient components of the ancestral and introduced fruits taken by ʻAlala? Do the rarer fruits (especially the ones that were common in their diet, e.g., ʻieʻie) contain elements, protein, or caloric value not replaced by those now more common? Do the introduced fruits contain elements (e.g., too much iron) that could cause long-term disorders?

Social behavior

We strongly discourage capturing adult birds for color-banding until their numbers increase. Initially, all data should be gathered through careful, noninvasive field observations. Like all corvids, the ʻAlala clearly can tolerate nearby humans (and cattle) engaged in nonthreatening behavior. They can easily tolerate noninvasive observations made away from the nest once a complete clutch is laid; however, capture and banding activities might produce unnecessary risk and stress. Scrutiny of the few remaining wild birds will provide clues for identifying many of them without color bands. Tape recordings should be analyzed to determine whether voice imprints can identify individual birds. Also, individual birds can possibly be tracked by documenting wing molt patterns.

All 'Alala newly released into the wild should be banded, and these birds will begin to provide additional data on social and dispersal behavior. Some released birds should be equipped with radiotelemetry transmitters that are glued to back or rump feathers and designed to fall off with the first annual molt. These will provide data on movement patterns by nonbreeders and might also provide the best opportunity to ascertain causes of death in this age class.

The role of prebreeding 'Alala in family affairs, especially in cooperative announcement or defense of territories, should be studied, as should the nature and duration of the lingering association between parents and nonbreeding offspring.

The nature of dispersal forays by prebreeders should be studied, as well as the age at which they begin, how far they move, whether dispersers regularly return to the natal family group, and to what habitats and elevations dispersing nonbreeders are attracted.

As to the degree of territorial aggression during the nonbreeding season, do family groups (including adjacent territorial pairs) merge, or are the nonbreeding "flocks" merely assemblages of different age classes of prebreeding 'Alala within a single family group?

Demography

Some of the most important demographic data are impossible to obtain from an unbanded population, especially a small population restricted to a forested habitat. Close study of the 'Alala, however, might provide clues relevant to the following unanswered questions:

- The causes of adult mortality.
- The hatchability of eggs in the first and subsequent clutches.
- The causes of brood reduction after hatching.
- The age of first breeding needs to be established for future demographic analyses. Although it is clear that some birds breed when 2 years old, the average age in males and females and the range of individual variation in age of sexual maturity need to be established.

The Captive Population

Findings

The husbandry and management of the captive 'Alala at Olinda are inadequate, and there is room for the modification and improvement of methods and procedures.

The age structure of the captive population of 10 birds is unstable and the pedigree is relatively shallow.

FINDINGS AND RECOMMENDATIONS

The captive population needs to be expanded as rapidly as possible for demographic and genetic purposes.

Incompatibility exists with some of the current pairings of the captive birds.

The Captive-Breeding Facility

Recommendation 11: ***A facility should be built that provides an optimal environment for successful breeding of the 'Alala. One husbandry facility is not adequate.***

The current facility is not optimal for successful breeding of the 'Alala. Characteristics of an optimal facility (or facilities) would include the following:

- An up-to-date library at the Olinda Facility, with continuing access to other journals and information from zoos on the mainland, including videos on various aspects of husbandry.
- Adequate veterinary facilities at Olinda, institutionalization of routine and emergency veterinary care, and a consistent pathology program.
- Additional crow enclosures that incorporate the design recommendations of experienced aviculturists from other captive-propagation programs and that allow for the safe and rapid separation of males from their mates.
- Increase the frequency and duration of routine daily activities around the birds to minimize the stress through conditioning. This process should be initiated during the non-breeding season.
- Long-term arrangements for pathological studies with a board-certified pathologist or pathology center.

The following specific features can be instituted immediately:

- Separate areas for juvenile and adults birds (existing structures can be used for juveniles).
- "Play" items in the aviaries--e.g., bark, games, or a "reward box"--to reduce aggressive behavior.
- More than one feeding area in each pen.
- A more natural environment in the form of planted aviaries, rather than wooden planters.

The following specific features require planning, cooperation, and coordination of several parties:

- A second captive-breeding facility should be developed to provide increased security from disease or other catastrophes. This second facility should be developed in Hawai'i and its collection should be restricted to endemic Hawaiian species.

● Advice from outside experts about the design of an optimal habitat for the 'Alala. A "natural" habitat might be produced with dirt floors and vines as a natural barrier between sections of the aviary; wire, which can cause injury to the birds, should be used only when necessary.

● Plexiglass and larger-meshed cages to provide more light for the birds. Increased light will also allow vegetation to grow in the aviaries and enhance the health of the birds.

● Changes in dimensions of the pens. The pens should be lower (10-12 ft. high) and wider, so that the birds can be captured with a minimum of trauma and risk of injury.

● New juvenile cages that include flight areas.

● New breeding pens for adult breeders with a configuration more suitable for nesting, including multiple nest baskets. Breeding enclosures for individual pairs should be physically separated by about 50 meters, if possible; otherwise, visual barriers should be installed on the existing breeding pens (perhaps with tennis netting and vegetation) before the breeding season.

Husbandry in the Captive-Breeding Program

Recommendation 12: ***Husbandry at the current captive breeding program must be improved.***

The following actions should be instituted immediately:

● Re-mate birds that have been unsuccessfully paired for more than 2 years.

● Check semen quality of males and relate it to known history of avian pox infection.

● Modify breeding enclosures so that male birds can be removed from the nesting site as soon as mates lay their first eggs.

● Allow females to incubate eggs for 5-7 days before the eggs are removed for artificial incubation, and reintroduce males to their mates only after clutch removal.

● Artificially incubate eggs at a dry-bulb temperature of 100.5°F. Monitor weight loss of each egg and adjust conditions to achieve correct loss to pip (about 12%).

● If necessary, radiograph eggs at the end of incubation to determine whether chicks are malpositioned within eggs.

● Hatch chicks under still-air conditions and use behavioral stimulation (tactile and auditory) to encourage weak chicks to hatch. Break out weak chicks as soon as possible during the pip-to-hatch interval (after the chorioallantoic blood system has shut down).

● Monitor nutritional state for components required for reproduction; e.g., check for normal blood concentrations of vitamin E, biotin, iron, and copper. Feed birds a more frugivorous, low-acid, low-iron diet.

The following actions require training, planning, cooperation, and coordination:

● Develop artificial-insemination techniques using another *Corvus* species as an appropriate research surrogate.

• Set up more than one captive-breeding facility.

• Augment the captive populations until there is a minimum of 40 breeding pairs (plus the number of non-breeders and subadults necessary to maintain this number) at two or more captive facilities. This "target" population size should be adjusted upward or downward depending on the effectiveness of other management strategies for increasing the size of the wild population. Any additions to the captive stock should be made first by taking first clutches of wild eggs and incubating them in the laboratory, second, by taking nestlings or recent fledglings, and third by taking some nonbreeding adults. Removal of adults from the wild should only be considered as a last resort.

• Develop chick-rearing protocols that minimize imprinting and maximize independence.

• Develop and implement safe and effective quarantine and sanitation procedures and protocols.

Data

Recommendation 13: ***More data on eggs, chicks, fledglings, and adults are needed to determine the causes of the lack of reproductive success of the captive population of 'Alala.***

General data

• A computerized record-keeping system that uses ISIS/ARKS and ISIS/SPARKS and an auxiliary database to track survival of eggs and chicks.

• ISIS/SPARKS to monitor average level of inbreeding.

• Systematic record-keeping, including information on breeding, fate of clutches, and egg quality.

• Consistent evaluation of embryo and bird mortality by qualified personnel. This will require accurate record keeping for such information as the fate of eggs and artificial incubation.

• Standard and regular genetic and demographic analyses on the captive population to establish annual objectives and pairing recommendations.

• Development of protocols for standard and nonstandard procedures, their wide circulation for peer review, and their implementation.

• Protocols for pathology.

Specific data

• Vitamins, minerals, amino acids and essential chemicals that are above or below normal concentrations and that could be adversely influencing sexual behavior, reproductive physiology, fertility, hatchability of eggs, or viability of young, with emphasis on vitamin E, biotin, iron, copper, and tannins.

• Comparison of the reproductive condition of nonbreeding adults and active breeders, based on direct laparoscopic examination or monitoring of hormone concentrations, to determine why some adults do not breed in captivity and others do.

● Optimal conditions for artificial incubation of crow eggs. Factors to consider are temperature, humidity, turning rate, still-air versus forced-air incubation in the first 7 days of incubation, and starting incubation under the female parent to improve hatchability.

● The feasibility of using human-imprinted crows for artificial insemination in captive propagation. Initial experimentation should be accomplished using another Corvus species as a research surrogate. Research surrogates for this work should not be housed at Olinda or any other facility housing ‘Alala.

● Sexual and social compatibility of all pairings.

● The effect of photoperiodicity on the timing of reproduction in crows.

● Other environmental stimuli that are necessary for reproduction, e.g., special foods for courtship feeding, perches for copulation, vocalizations, stimuli from adjacent pairs, and external stimuli that inhibit reproduction.

Adults

● DNA analyses of all captive ‘Alala to document the extent of band-sharing.

● Chromosome analysis of all captive ‘Alala to look for chromosomal abnormalities.

● Electrophoretic analysis of proteins for all captive ‘Alala to determine the extent of heterozygosity.

● Development of a complete protocol for behavioral observation, including television monitors and a blind. Information gathered would help to determine when first eggs are laid so that male birds can be removed from their breeding enclosures promptly.

● Measurements of what the birds are eating, as opposed to what they are fed, to determine nutritional needs, including measurement of circulating vitamins through analysis of blood samples.

● Weights of captive birds over time.

● Determination of hematological and serum norms to assist in the diagnosis of illness.

● Yearly veterinary evaluations, including blood sampling and screening for parasites.

● Observation and description of molting of birds to determine potential toxicoses and sex of the birds. Analysis of molting patterns might also help to determine the age of birds in the wild.

● Sex of birds according to laparoscopy performed by a qualified avian veterinarian and application of chromosomal techniques to feathers.

● Institution of artificial insemination techniques in the captive population of ‘Alala with semen from a male nonbreeder and an imprinted female if initial breeding attempts are unsuccessful or result in the production of infertile eggs.

Chicks

● Development of a complete protocol for behavioral observation of chicks that emphasizes the avoidance of imprinting while maximizing independence, placing of young birds with other captive crows as soon as possible, and allowing them to pick their own food as soon as possible.

- Documentation of physical and behavioral development, including growth curves, molt sequences, and food intakes.
- Veterinary information.

Eggs

- Egg pathology, including bacteriological findings on all unsuccessfully hatched eggs.
- Bacterial culturing of dead embryos.
- Complete embryonic analysis.

Nutrition

Recommendation 14: ***Changes in the captive 'Alala's diet are needed.***

Although there are many uncertainties and unanswered questions about iron-storage disease, changes in the diet of captive 'Alala are warranted. A low-iron, low-acid balanced diet is recommended.

An accurate nutritional analysis of food items in the wild is needed, and consideration should be given to simplifying the diet in captivity to ensure proper and consistent nutrition.

Equipment at the Captive-Breeding Facility

Recommendation 15: ***Equipment should be modernized to ensure the success of the captive-breeding program.***

The following equipment is needed to support a successful captive-breeding program:

- A full veterinary laboratory-clinic (equivalent to the facilities of a small-animal practice), including quarantine areas, an x-ray machine, updated video-monitoring equipment, a library, and a hatchery (with a hatching room and a brooding facility).
- A pathology laboratory somewhere on the Hawaiian islands.

Personnel and Training

Recommendation 16: ***A mechanism must be established for continuing training of personnel at the captive-breeding facility.***

Adequately trained personnel are essential to the success of the captive-breeding program. To correct current shortcomings, enhancement activities should include the following:

- A fully funded, long-term avicultural training program for staff at Olinda, including periodic visits to mainland zoos and other captive breeding facilities for workshops and additional training and programs tailored to the specific needs of trainees and the breeding facility.
- Training in all aspects of genetic management of small populations, including the use of computer software designed to provide those analyses.
- An apprenticeship program.
- Access to state-of-the-art technologies developed at zoos and other captive breeding facilities.

The facility staff should include the following:

- A full-time director who serves as both a curator and an administrator. The person should be knowledgeable about both aviculture and ornithology and should be up to date on avicultural techniques. The person would be responsible for the overall management of the facilities, including training of keeper staff, time management, administration, public relations, and fund-raising.
- A full-time on-site avian veterinarian.
- A full-time on-site aviculturist.
- At least two additional animal-keepers.

Genetics of the Wild and Captive Populations

Findings

The decline of the 'Alala must have been accompanied by a loss of genetic variation as a result of genetic drift and inbreeding, but the extent of this loss is unknown. Inbreeding itself is unlikely to be a major cause of the decline of the 'Alala in the wild. Lack of molecular data about historical and current levels of genetic variation in the wild population have prevented meaningful investigation of the genetic implications of the 'Alala's decline. In any case, the 'Alala population on the McCandless Ranch is so close to the wild sources of some of the captive birds that they would be expected to add no more than very slight genetic variability to the captive population, and even the current magnitude of variation will be retained only if the population is rapidly increased.

Assuming that all founders of the captive population were originally unrelated (which might or might not be the case), eight of the surviving birds have calculated inbreeding coefficients of 0, and two (the female Hooku and the male Hoikei) have inbreeding coefficients of 0.25. More importantly, two of the four current breeding pairs will produce offspring with inbreeding coefficients of 0.25.

FINDINGS AND RECOMMENDATIONS

DNA fingerprinting using two probes reveals a high proportion of band-sharing for seven captive 'Alala. A comparison of data from three other wild corvid populations suggests that the captive population is strongly inbred. The absence of comparable data from the wild 'Alala, however, limits the usefulness of these data. It is not clear whether the extant wild population is suffering from inbreeding depression caused by the loss of genetic variation.

Recommendation 17: ***Because the wild population is so small and the need for increasing numbers of birds is so critical, addition of new wild-caught adult birds to the captive stock should have a very low priority until the wild population has increased, because this action is not likely to provide a genetic advantage.***

Release of captive birds on the McCandless Ranch solely for the purpose of augmenting genetic variability has no supporting rationale. Any releases must be part of a full-scale management plan.

At the present time, genetic studies on the wild population should be considered to be of secondary importance. It is clear that the preservation of genetic diversity in both the captive and wild populations will require that these populations be increased in size as rapidly as possible. Until the wild and captive populations increase substantially, demographic considerations should remain the principal determinants of management activities. However, genetic analysis (e.g., DNA, electrophoresis, and chromosome analyses) should be completed on all captive birds, and on all offspring of wild pairs that are subject to manipulation in any form. This information could prove especially valuable for recovery management, but if not, we have still performed an important function of documenting data that could prove useful in future conservation efforts with other species.

REFERENCES

Allendorf, F. W., and R. F. Leary. 1986. Heterozygosity and fitness in natural populations of animals. Pp. 57-76 in M. E. Soulé, ed. Conservation Biology: The Science of Scarcity and Diversity. Sunderland, Mass: Sinauer Assocs.

Anderegg, R., H. Frey, and H. U. Miller. 1983. Reintroduction of the bearded vulture or Lammergeier *Gypaetus barbatus aureus* to the Alps. Int. Zoo Yearb. 23:35-41.

Andrewartha, H. G., and L. C. Birch. 1986. The Ecological Web: More on the Distribution and Abundance of Animals. Chicago: Chicago University Press. 506 pp.

Anon. 1988. A new home for the Seychelles Brush Warbler. World Birdwatch 10 (3,4):4.

Anon. 1991a. Overview, New Zealand. Am. Birds, Winter 1991:1040.

Anon. 1991b. Success for takahe. Oryx 25:14.

Anon. 1991c. Recent news from Bali. World Birdwatch 13(2):3.

Antikanen, E. 1978. The breeding adaptations of the Jackdaw *Corvus monedula L.* in Finland. Savonia 2:1-45.

Atkinson, I. A. E. 1970. Successional trends in the coastal and lowland forest of Mauna Loa and Kilauea volcanoes, Hawaii. Pacific Sci. 24:387-220.

Atwood, J. L., M. J. Elpers, and C. T. Collins. 1990. Survival of breeders in Santa Cruz Island and mainland California Scrub Jay populations. Condor 92:783-788.

Baldwin, P. H. 1969. The Alala (*Corvus tropicus*) of western Hawaii Island. 'Elepaio 30: 41-45.

Baldwin, P.H., and G.O. Fagerlund. 1943. The effect of cattle grazing on koa reproduction in Hawaii National Park. Ecology 24:118-122.

REFERENCES

Ballou, J. D. 1991. Management of genetic variation in captive populations. Pp. 602-610 in E. C. Dudley, ed. The Unity of Evolutionary Biology: Proceedings of the Fourth Interantional Congress of Systematic and Evolutionary Biology. Portland, Oregon: Dioscorides Press. [University of Maryland, College Park, USA] July 1990.

Banko, P. C. 1974. Report on the 'Alala (*Corvus tropicus*). Unpubl. report submitted to World Wildlife Fund and Hawaii State Division of Fish and Game. 31 pp.

Banko, P. C. 1976. Saving the 'Alala (*Corvus tropicus*) -- some preliminary recommendations. Pp. 23-28 in C. W. Smith, ed. Proceedings of the First Conference in Natural Sciences in Hawaii, held at Hawaii Field Research Center, Hawaii Volcanoes National Park, August 19-20, 1976. Honolulu: University of Hawaii.

Banko, W. E., and P. C. Banko. 1980. History of endemic Hawaiian birds. Part 1. Population histories - species accounts: Forest birds: Alala or Hawaiian Raven/Crow. Avian Hist. Rep. 6B. Coop. Natl. Park Resour. Stud. Unit, University of Hawaii, Manoa, Honolulu, 121 pp.

Bent, A. C. 1946. Life histories of North American jays, crows, and titmice. U.S. Natl. Mus. Bull. 191:1-494.

Berger, A. J. 1981. Hawaiian birdlife. 2nd Ed. Honolulu: University of Hawaii Press.

Blackwell, A. 1990. The White Stork: A capsule overview of its status in the wild and the role of captive reared and bred birds in re-introduction. Avic. Mag. 96:105-114.

Bloesch, M. 1980. Drei Jahrzehnte Schweizerischer Storchansied-lungsversuch *Ciconia ciconia* in Altreu, 1948-1979. Ornithologische Beob. 77:167-194.

Bluhm, C. K. 1985. Mate preferences and mating patterns of Canvasbacks (*Aythya valisineria*). Ornithol. Monogr. 37:45-56.

Brock, M. K. 1991. A Comparison of DNA Fingerprints from Hispaniolan and Puerto Rican Parrots. El Pittirre 4(3):3-4.

Broo, B. 1978. Project Eagle Owl, Southwest. Pp. 104-120 in T. A. Greer, ed. Birds of Prey Management Techniques. Oxford: British Falconers Club.

Brown, J. L., and E. R. Brown. 1990. Mexican Jays: uncooperative breeding. Pp. 269-288 in Stacey, P. B. and W. P. Koenig, eds. Cooperative Breeding in Birds: Long-Term Studies of Ecology and Behavior. Cambridge: Cambridge University Press.

Bryant, I. J. 1985. Rare birds in National Wildlife Centre, Mt. Bruce, New Zealand. Avic. Mag. 91:48-59.

Buitron, D. 1988. Female and male specialization in parental care and its consequences in Black-billed Magpies. Condor 90:29-39.

Bulmer, M. G. 1973. Inbreeding in the Great Tit. Heredity 30:313-325.

Burley, N., and N. Moran. 1979. The significance of age and reproductive experience in the mate preferences of feral Pigeons, *Columba livia*. Anim. Behav. 27:686-698.

Burr, T. A. 1984. Alala Restoration Plan. Dept. of Land and Natural Resources, Division of Forestry and Wildlife, Honolulu, Hawaii.

Burr, T. A., P. Q. Tomich, E. Kosaka, W. Kramer, J. M. Scott, E. Kridler, J. Giffin, D. Woodside, and R. Bachman. 1982. Alala Recovery Plan. Portland, Oregon: U.S. Fish and Wildlife Service.

Butler, R. W., N. A. M. Verbeek, and H. Richardson. 1984. The breeding biology of the Northwestern Crow. Wilson Bull. 96:408-418.

Butler, P. 1992. Parrots, pressures, people, and pride. Pp. 25-46 in S. R. Beissinger and N. F. R. Snyder, eds. New World Parrots in Crisis: Solutions from Conservation Biology. Washington, D.C.: Smithsonian Institution Press.

Cade, T. J. 1986a. Reintroduction as a method of conservation. Raptor Res. 5:72-84.

Cade, T. J. 1986b. Propagating diurnal raptors in captivity: A review. Int. Zoo Yearb. 24/25:1-20.

Cade, T. J. 1988. Using science and technology to reestablish species lost in nature. Pp. 279-288 in E. O. Wilson, ed. Biodiversity. Washington, D.C.: National Academy Press.

Cade, T. J. 1990. Peregrine falcon recovery. End. Species Update 8:40-43.

Cade, T. J., and V. J. Hardaswick. 1985. Summary of Peregrine Falcon production and reintroduction by the Peregrine Fund in the United States, 1973-1984. Avic. Mag. 91:79-92.

Cade, T. J., and C. G. Jones. In Press. Progress in restoration of the Mauritius Kestrel.

REFERENCES

Cade, T. J., J. H. Enderson, C. G. Thelander, and C. M. White, eds. 1988. Peregrine Falcon Populations: Their Management and Recovery. Boise, Idaho: Peregrine Fund, Inc.

Carpenter, J. W., and S. R. Derrickson. 1981. The role of captive propagation in preserving endangered species. Pp. 109-113 in R. R. Odom and J. W. Guthrie, eds. Proceedings of the Nongame and Endangered Wildlife Symposium, August 13-14, 1981. Athens, Georgia, Georgia Dept. of Natural Resources, Game and Fish Division. Technical Bulletin WL 5.

Carpenter, J. W., R. R. Gabel, and J. G. Goodwin, Jr. 1991. Captive breeding and reintroduction of the endangered Masked Bobwhite. Zoo Biol. 10:439-450.

Carson, H. L. 1990. Increased genetic variance after a population bottleneck. Trends Ecol. Evol. 5:228-230.

Carson, H. L., J. P. Lockwood, and E. M. Craddock. 1990. Extinction and recolonization of local populations on a growing shield volcano. Proc. Natl. Acad. Sci. USA 87:7055-7057.

Cavill, J. P. 1982. Viral diseases. Pp. 515-527 in Nil Petrak, ed. Diseases of Cage and Aviary Birds. Philadelphia: Lea and Febiger.

Chamberlain-Auger, J. A., P. J. Auger, and E. G. Strauss. 1990. Breeding biology of American Crows. Wilson Bull. 102:615-622.

Clapp, R. B., M. K. Klimkiewicz, and A. G. Futcher. 1983. Longevity records of North American birds: Columbidae through Paridae. J. Field Ornithol. 54:123-137.

Clark, T. W. 1990. Black-footed ferrets on the road to recovery. End. Species Update 8:86-88.

Clubb, K. J. 1991. The reintroduction of Military Macaws in Guatemala: An example of private aviculture's role in avian conservation. PsittaScene 3:2-3.

Coles, D. 1978. Breeding the Hunting Cissa at Padstow Bird Gardens. Avic. Mag. 84: 183-185.

Coles, D. 1980. Breeding the Plush-Capped Jay at Padstow Bird Gardens. Avic. Mag. 86:125-128.

Collar, N. J., and P. Andrew. 1988. Birds to Watch: The ICBP World Checklist of Threatened Birds. ICBP Tech. Publ. No. 8. Washington, D.C.: Smithsonian Institution Press.

Conant, S. 1988. Saving endangered species by translocation: Are we tinkering with evolution? Bio Science 38:254-257

Conway, W. 1988. Can technology aid species preservation? Pp. 263-268 in E.O. Wilson, ed. Biodiversity. Washington, D.C.: National Academy Press.

Coombs, F. 1978. The Crows. A Study of the Corvids of Europe. London: Batsford.

Cox, P. A. 1983. Extinction of the Hawaiian avifauna resulted in a change of pollinators for the ieie, *Freycinetia arborea*. Oikos 41:195-199.

Craig, J. L. 1991. Are small populations viable? ACTA XX Congressus Internationis Ornithologica 4:2546-2552. 2-9 December 1990, Christchurch, New Zealnd. Wellington: New Zealand Ornithological Congress Trust Board.

Daniell, A., and N. D. Murray. 1986. Effects of inbreeding in the Budgerigar *Melopsittacus undulatus* (Aves:Psittacidae). Zoo Biol. 5:233-238.

Davies, N. T. 1979. Anti-nutrient factors affecting mineral utilization. Proc. Nutr. Soc. 38:121-128.

Delacour, D. J. 1936. Aviculture. Vol 1. Hertford: Stephen Austin.

Denniston, C. 1977. Small population size and genetic diversity: Implications for endangered species. Pp. 281-289 in S. A. Temple, ed. Endangered Birds: Management Techniques for Preserving Threatened Species. Madison: University of Wisconsin Press.

Derrickson, S. R. 1985. Captive propagation of whooping cranes, 1982-1984. Pp. 377-386 in J. C. Lewis, ed. Proceedings 1985 Crane Workshop. Grand Island, Nebraska: Platte River Whooping Crane Habitat Maintenance Trust.

Derrickson, S. R., and J. W. Carpenter. 1987. Behavioral management of captive cranes -- factors influencing propagation and reintroduction. Pp. 493-514 in G.W. Archibald and R.F. Pasquier, eds. Proceedings 1983 International Crane Workshop. Baraboo, Wisconsin: International Crane Foundation.

REFERENCES

Derrickson, S. R., and N. F. R. Snyder. 1992. Potentials and limits of captive breeding in parrot conservation. Pp. 133-163 in S. R. Beissinger and N. F. R. Snyder, eds. New World Parrots in Crisis: Solutions from Conservation Biology. Washington, D.C.: Smithsonian Institution Press.

Dhindsa, M. S., and D. A. Boag. 1990. The effect of food supplementation on the reproductive success of Black-billed Magpies *Pica pica*. Ibis 132: 595-602.

Diamond, J. M. 1985. Population processes in island birds: Immigration, extinction and fluctuations. Pp. 17-21 in International Council for Bird Preservation. ICBP Tech. Publ. No. 3.

Drewien, R. C., and E. G. Bizeau. 1977. Cross-fostering whooping cranes to sandhill crane foster parents. Pp. 201-222 in S.A. Temple, ed. Endangered Birds: Management Techniques for Preserving Threatened Species. Madison: University of Wisconsin Press.

Durrell, L., and J. Mallinson. 1987. Reintroduction as a political and educational tool for conservation. Dodo 24: 6-19.

Duvall, F., C. Tarr, and R. Fleischer. 1991. Genetic variation and inbreeding in the severely endangered Hawaiian Crow or "Alala" - a bird at a bottleneck. Poster presented at the meeting: Hawaiian Evolution 1991, University of Hawaii at Hilo.

Edworthy, T. 1972. Breeding Choughs at the Paignton Zoo. Int. Zoo Yearb. 12:140-141.

Ellis, D. H., S. J. Dobrott, and J. B. Goodwin, Jr. 1977. Reintroduction techniques for Masked Bobwhites. Pp. 345-354 in S.A. Temple, ed. Endangered Birds: Management Techniques for Preserving Threatened Species. Madison: University of Wisconsin Press.

Emlen, J. T., Jr. 1942. Notes on a nesting colony of western crows. Bird-Banding 13:143-154.

Engbring, J. 1990. U.S. Fish and Wildlife Service Memorandum, July 31.

Evendon, F. G., Jr. 1947. Nesting studies of the Black-billed Magpie in southern Idaho. Auk. 64:260-266.

Fairbanks, V. F., J. L. Fahey, and E. Beutler. 1971. Clinical Disorders of Iron Metabolism. New York: Grune and Stratton, Inc.

Fentzloff, C. 1984. Breeding, artificial incubation and release of White-tailed Sea Eagles *Haliaeetus albicilla*. Int. Zoo Yearb. 23:18-35.

Fitzpatrick, J. W., and G. E. Woolfenden. 1986. Demographic routes to cooperative breeding in some New World jays. Pp. 137-160 in Nitecki, M. and J. Kitchell, eds. Evolution of Behavior. Chicago: University of Chicago Press.

Flack, J. A. D. 1977. Interisland transfers of New Zealand Black Robins. Pp. 365-372 in S. A. Temple, ed. Endangered Birds: Management Techniques for Preserving Threatened Species. Madison: University of Wisconsin Press.

Frankel, O. H. and M. E. Soulé. 1981. Conservation and Evolution. New York: Cambridge University Press.

Franklin, I. R. 1980. Evolutionary changes in small populations. Pp. 135-149 in M. E. Soulé and B. A. Wilcox, eds. Conservation Biology: An Evolutionary-Ecological Perspective. Sunderland, Massachusetts: Sinauer Assocs.

Fullagar, P. J. 1985. The woodhens of Lord Howe Island. Avic. Mag. 91:15-30.

Garcelon, D. K., and G. W. Roemer, eds. 1988. Proceedings of the International Symposium on Raptor Reintroduction, 1985. Arcata, California: Institute for Wildlife Studies.

Gretton, A., J. Komdeur, and M. Komdeur. 1991. Saving the Magpie Robin. World Birdwatch 13:10-11.

Gibson, L. 1980. Steller's Jay, *Cyanocitta stelleri*. Avic. Mag. 86:141-147.

Giffin, J. G. 1983. 'Alala investigation final report. Pittman-Robertson Proj. W-18-R, Study R-IIB, 1976-1981. Dept. of Land and Natural Resources, Division of Forestry and Wildlife, Honolulu, Hawaii.

Giffin, J. G., J. M. Scott, and S. Mountainspring. 1987. Habitat selection and management of the Hawaiian Crow. J. Wildl. Manage. 51:485-494.

Giffin, J. G. 1990. Limited surveys of forest birds and their habitats in the state of Hawaii; Puu Waawaa Wildlife Sanctuary Bird Survey, July 1, 1989 to June 30, 1990. Unpubl. Prog. Rep., Proj. No W-18-R-15. Honolulu: Hawaii Dept. of Land and Nat. Resources. 8 pp.

REFERENCES

Giffin, J. G. 1991. Limited surveys of forest birds and their habitats in the state of Hawaii; Puu Waawaa Wildlife Sanctuary Bird Survey, July 1, 1990 to June 30, 1991. Unpubl. Prog. Rep., Proj. No W-18-R-16. Honolulu: Hawaii Dept. of Land and Nat. Resources. 2 pp.

Goodburn, S. F. 1991. Territory quality or bird quality? Factors determining breeding success in the Magpie *Pica pica*. Ibis 133:85-90.

Goodman, D. 1987. The demography of chance extinction. Pp. 11-34 in M. E. Soulé, ed. Viable Populations for Conservation. New York: Cambridge University Press.

Goodwin, D. 1954. Lanceolated Jays breeding in captivity. Avic. Mag. 60:154-163.

Goodwin, D. 1987. Update Crows of the World. Seattle: University of Washington Press.

Grahame, I. 1980. Re-introduction of captive bred Cheer Pheasants *Catreus wallichi*. Int. Zoo Yearb. 20:36-40.

Grahame, I. 1988. Up-date on the World Pheasant Association Cheer Pheasant re-introduction project in Pakistan. Pp. 615-622 in B. L. Dresser, R. W. Reece and E. J. Maruska, eds. Proceedings 5th World Conference on Breeding Endangered Species in Captivity. Cincinnati Zoo and Botanical Garden Cincinnati, Ohio: Cincinnati Zoo. October 9-12, 1988.

Greenlaw, J. S., and R. F. Miller. 1983. Calculating incubation periods of species that sometimes neglect their last eggs: The case of the Sora. Wilson Bull. 95:459-461.

Greenwell, G. A., C. Emerick, and M. Biben. 1982. Inbreeding depression in Mandarin Ducks: A preliminary report on some continuing experiments. Avic. Mag. 88:145-148.

Greenwood, P.J. 1980. Mating systems, philopatry and dispersal in birds and mammals. Anim. Behav. 28:1140-1162.

Grier, J. W., and R. W. Fyfe. 1987. Preventing research and management disturbance. Pp. 173-182 in B. A. Giron Pendleton, B. A. Millsap, K. W. Cline, and D. M. Bird, eds. Raptor Management Techniques Manual. Washington, D.C.: National Wildlife Federation.

Griffith, B., J. M. Scott, J. W. Carpenter, and C. Reed. 1989. Translocation as a species conservation tool: Status and strategy. Science 245:477-480.

Gwinner, E. 1965. Beobachtungen uber Nestbau und Brutpflege des Kolkraben (*Corvus corax*) in Gefangenschaft. J. Ornithol. 106:145-178.

Haig, S. M., J. D. Ballou, and S. R. Derrickson. 1990. Management options for preserving genetic diversity: Reintroduction of Guam Rails to the wild. Conserv. Biol. 4:290-300.

Hay, R. 1990. New Zealand: The turning point. World Birdwatch 12:11.

Hedrick. P. W., and P. S. Miller. 1992. Conservation genetics: techniques and fundamentals. Ecol. Applications 2:30-46.

Henshaw, H. W. 1902. Birds of the Hawaiian Islands, being a complete list of the birds of the Hawaiian possessions with notes on their habits. Honolulu: T. G. Thrum.

Hochachka, W. M. 1988. The effect of food supply on the composition of Black-billed Magpie eggs. Can. J. Zool. 66: 692-695.

Hochachka, W. M., and D. A. Boag. 1987. Food shortage for breeding Black-billed Magpies (*Pica pica*): An experiment using supplemental food. Can. J. Zool. 65:1270-1274.

Holyoak, D. 1967. Breeding biology of the Corvidae. Bird Study 14: 153-168.

Hoshide, H. M., A. J. Price, and L. Katahira. 1990. A progress report on Nene *Branta sandvicensis* in Hawaii Volcanoes National Park from 1974-89. Wildfowl 41:152-155.

Husby, M. 1986. On the adaptive value of brood reduction in birds: Experiments with the Magpie *Pica pica*. J. Anim. Ecol. 55:75-83.

International Species Inventory System (ISIS). 1984-1991. Species Distribution Report Abstracts -- Birds. Apple Valley, Minnesota: ISIS.

ISIS. 1990. ARKS: Animal Records Keeping System. Apple Valley, Minnesota: International Species Inventory System.

International Union for the Conservation of Nature (IUCN). 1990. The 1990 IUCN Red List of Threatened Animals. Gland, Switzerland: International Union for the Conservation of Nature. 228 pp.

REFERENCES

International Zoo Yearbook. 1959-1990. Species of birds bred in captivity and multiple generation births. Int. Zoo Yearb. 1:156; 2:277; 3:283; 4:254; 5:364; 6:424; 7:347; 8:338; 9:267; 10:304; 11:315-316; 12:371; 13:315; 14:363-364; 15:355; 16:375; 17:296; 18:365; 19:344; 20:415; 21:304; 22:405; 23:307; 24/25:505-506; 26:466; 27:395; 28:444; 29:305-306.

James, H. F., and S. L. Olson. 1991. Descriptions of thirty-two new species of birds from the Hawaiian Islands: Part II. Passeriformes. Ornithol. Monogr. 46: 1-88.

Jenkins, C. D., S. A. Temple, C. van Riper, and W. R. Hansen. 1989. Disease-related aspects of conserving the endangered Hawaiian Crow. ICBP Tech. Publ. 10:77-87.

Jones, C. D., and A. W. Owadally. 1988. The life histories and conservation of the Mauritius Kestrel *Falco punctatus* (Temminck 1823), Pink Pigeon *Columba mayeri* (Prevost 1843), and Echo Parakeet *Psittacula eques* (Boddaert 1783). Proc. Royal Soc. of Arts & Sci. of Mauritius 5:80-130.

Jones, C. G., F. N. Steele, and A. W. Owadally. 1981. An account of the Mauritius Kestrel captive breeding project. Avic. Mag. 87:191-207.

Jones, C. G., D. M. Todd, K. J. Swinnerton, and Y. Mungroo. 1988. The release and management at liberty of captive bred Pink Pigeons *Columba mayeri* on Mauritius. Pp. 393-414 in B. L. Dresser, R. W. Reece, and E. J. Maruska, eds. Proceedings 5th World Conference on Breeding Endangered Species in Captivity, October 9-12. Cincinnati, Ohio: Cincinnati Zoo and Botanical Garden.

Jones, C. G., D. M. Todd, and Y. Mungroo. 1989. Mortality, morbidity and breeding success of the Pink Pigeon (*Columba (Nesoena) mayeri*). Pp. 89-113 in J. E. Cooper, ed. Disease and Threatened Birds: ICBP Tech, Publ. No. 10. Cambridge: International Council for Bird Preservation.

Jones, C. G., W. Heck, R. Lewis, Y. Mungroo, and T. J. Cade. 1991. A summary of the conservation management of the Mauritius Kestrel (*Falco punctatus*).

Kear, J. 1986. Captive breeding programs for waterfowl and flamingos. Int. Zoo Yearb. 24/25:21-25.

Kear, J., and A. J. Berger. 1980. The Hawaiian Goose. Calton, Poyser.

Kepler, C. B. 1977. Captive propagation of Whooping Cranes: a behavioral approach. Pp. 231-241 in S. A. Temple, ed. Endangered Birds: Management Techniques for Preserving Endangered Species. Madison: University of Wisconsin Press.

Kiff, L. California Condor Recovery Team Minutes - 16 October 1986. Unpublished Report.

Kiff, L. California Condor Recovery Team Minutes - 6 June 1989. Unpublished Report.

Kilham, L. 1984a. Cooperative breeding of American Crows. J. Field Ornithol. 55:349-356.

Kilham, L. 1984b. Play-like behavior of American Crows. Fla. Field Natl. 12:33-36.

Kilham, L. 1984c. Intra- and extrapair copulatory behavior of American Crows. Wilson Bull. 96:716-717.

Kilham, L. 1985a. Behavior of American Crows in the early part of the breeding cycle. Fla. Field Natl. 13:25-31.

Kilham, L. 1985b. Territorial behavior of American Crows. Wilson Bull. 97:389-390.

Kilham, L. 1986. Renestings of American Crows in Florida and predation by raccoons. Fla. Field Natl. 14:21-23.

Kincaid, A. L., and M. K. Stoskopf. 1987. Passerine dietary iron overload syndrome. Zoo Biol. 6:79-88.

King, W. B. 1977-1979. Red data book 2, Aves. Morges, Switzerland: International Union for the Conservation of Nature and Natural Resources.

King, W. B. 1985. Island birds: Will the future repeat the past? Pp. 3-15 in P. J. Moors ed. Conservation of Island Birds. ICBP Tech. Publ. No. 3. Cambridge, International Council for Bird Preservation.

Kleiman, D. 1989. Reintroduction of captive mammals for conservation: Guidelines for reintroducing endangered species into the wild. BioScience 39 (3):152-160.

Klint, T., and M. Enquist. 1980. Pair formation and reproductive output in domestic pigeons. Behav. Processes. 6:67-72.

Kress, S. W. 1977. Establishing Atlantic Puffins at a former breeding site. Pp. 373-377 in S.A. Temple, ed. Endangered Birds: Management Techniques for Preserving Threatened Species. Madison: University of Wisconsin Press.

REFERENCES

Kuehler, C. M. 1989. California condor (*Gymnogyps claifornianus*) studybook. San Diego: San Diego Wild Animal Park. 25 pp.

Kuehler, C. M. 1992. 'Alala-Hawaiian Crow Studbook.

Kuehler, C. M., and J. Good. 1990. Artificial incubation of bird eggs at the Zoological Society of San Diego. Int. Zoo Yearb. 29:118-136.

Kuehler C. M., and P. Witman. 1988. Artificial incubation of California condor (*Gymnogyps californianus*) eggs removed from the wild. Zoo Biol. 7:123-132.

Kuehler, C. M., D. J. Sterner, D. S. Jones, R. L. Usnik, and S. Kasielke. 1991. Report on captive hatches of California Condors (*Gymnogyps californianus*): 1983-1990. Zoo Biol. 10:65-68.

Kuehler, C. M., A. Lieberman, B. McIlraith, W. Everett, T. A. Scott, M. L. Morrison, and C. Winchell. In press. Artificial incubation and hand-rearing of Loggerhead shrikes. J. Wildlife Management.

Lacy, R. C. 1987. Loss of genetic diversity from managed populations: Interacting effects of drift, mutation, immigration, selection and population subdivision. Conserv. Biol. 1:143-158.

Lacy, R. C. 1989. Analysis of founder representation in pedigrees: founder equivalents and founder genome equivalents. Zoo Biol. 8:111-123.

Lamm, D. W. 1958. A nesting study of the Pied Crow in Accra, Ghana. Ostrich 29:59-70.

Lande, R. 1988. Genetics and demography in biological conservation. Science 241:1455-1460.

Lawton, M. F., and C. F. Guinton. 1981. Flocking composition, breeding success, and learning in the Brown Jay. Condor 83:27-33.

Laycock, G. 1991. All-American survivor. Wildl. Conserv. 94:38-46.

Lerner, J. M. 1954. Genetic Homeostasis. New York: John Wiley & Sons.

Levins, R. 1970. Extinction. Pp. 77-107 in M. Gerstenhaber, ed. Some Mathematical Questions in Biology, Vol. II. Second Symposium on Mathematical Biology. Providence, Rhode Island: American Mathematical Society.

Lewin, V., and G. Lewin. 1984. The Kalij Pheasant, a newly established game bird on the island of Hawaii. Wilson Bull. 96:634-646.

Lewis, J. C. 1990. Captive propagation in the recovery of whooping cranes. End. Species Update 8:46-48.

Lieberman, A., J. W. Wiley, J. V. Rodriguez, and J. M. Paez. 1991. The first experimental reintroduction of captive-reared Andean condors *(Vultur gryphus)* into Columbia, South America. AAZPA Ann. Conf. Proc. 1991:129-136.

Ligon, J. D. 1971. Late summer-autumnal breeding of the Pinon Jay in New Mexico. Condor 73:147-153.

Lockie, J. D. 1955. The breeding and foraging of Jackdaws and Rooks with notes on Carrion Crows and other Corvidae. Ibis 97:341-369.

Lockwood, J. P., and P. W. Lipman. 1987. Holocene eruptive history of Mauna Loa volcano. U.S. Geological Survey Professional Paper 1350:500-535. Washington, D.C.: U.S. Government Printing Office.

Lourie-Fraser, G. 1983. Captive breeding of the Lord Howe Island Woodhen: an endangered rail. A.F.A. Watchbird 10:30-44.

Love, J.A. 1984. The Return of the Sea-Eagle. Cambridge: Cambridge University Press.

Lowenstein, L. J., and M. L. Petrak. 1978. Iron pigment in the livers of birds. Pp. 127-135 in R. J. Montali, and G. Migaki, eds. The Comparative Pathology of Zoo Animals. Washington, D.C.: Smithsonian Institution Press.

Lyles, A. M., and R. M. May. 1987. Problems leaving the ark. Nature 326:245-246.

Maguire, L. A. 1991. Risk analysis for conservation biologists. Conserv. Biol. 5:123-125.

McGowan, K. J., and G. E. Woolfenden. 1991. A sentinel system in the Florida Scrub Jay. Anim. Beha. 37:1000-1006.

McMillan, J. L., D. H. Ellis, and D. G. Smith. 1987. The role of captive propagation in the recovery of the Mississippi Sandhill Crane. End. Spec. Tech. Bull. 12:6-8.

Merton, D. V. 1975. The Saddleback...its status and conservation. Pp. 61-64 in R.D. Martin, ed. Breeding Endangered Species in Captivity. London: Academic Press.

REFERENCES

Merton, D., and R. Empson. 1989. But it doesn't look like a parrot. Birds International 1:60-72.

Miller, B., and H. F. Mullette. 1985. Rehabilitation of an endangered Australian bird: The Lord Howe Island Woodhen *Tricholimnas sylvestris* (Schlater). Biol. Conserv. 34:55-95.

Mills, J. A. and G. R. Williams. 1979. The status of New Zealand birds. Pp. 147-168 in M. Tyler, ed. Status of Australian Wildlife. Proceedings of the Centenary Symposium of the Royal Zoological Society of South Australia, 21-23 September, 1978.

Morner, T. 1986. Zoological gardens and the conservation of wildlife in Sweden. Int. Zoo Yearb. 24/25:189-192.

Morrison, M. 1991. San Clemente Loggerhead Shrike (*Lanius ludovicianus mearnsi):* Request to initiate captive propagation program. Unpubl. rept. to California Dept. of Game and Fish: U.S. Fish and Wildlife Service.

Moore, R. R., Clague, D. A., Rubin, M. and W. A. Borhrson. 1987. Hualalai Volcano: A preliminary summary of geologic, petrologic and geophysical data. U.S. Geological Survey Professional Paper 1350:571-585. Washington, D.C.: U.S. Government Printing Office.

Moorehouse, R. J., and R G. Powlesland. 1991. Aspects of the ecology of the Kakapo *Strigops habroptilus* liberated on Little Barrier Island (Huaturu), New Zealand. Biol. Conserv. 56:349-365.

Mueller-Dombois, D. 1987. Forest dynamics in Hawaii. Trends Ecol. and Evol. 2:216-220.

Mueller-Dombois, D., K. W. Bridges, and H. L. Carson, eds. 1981. Island Ecosystems: Biological Organization in Selected Hawaiian Communities. U.S. International Biological Program Series 15. Stroudsburg, Penn.: Hutchinson Ross Publ. Co. pp. 583.

Munro, G. C. 1944. Birds of Hawaii. Honolulu: Tongg.

Murphy, D. D., K. E. Freas, and S. B. Weiss. 1990. An environmental-metapopulation approach to population viability analyses for a threatened invertebrate. Conserv. Biol. 4:41-51.

Newton, I., P. E. Davis, and J. E. Davis. 1983. Ravens and buzzards in relation to sheep farming and forestry in Wales. J. Appl. Ecol. 19:681-706

Nice, M. M. 1954. Problems of incubation periods in North American birds. Condor 56:173-197.

Nye, P. E. 1988. A review of bald eagle hacking projects and early results in North America. Pp. 95-112 in D. K. Garcelon and G. W. Roemer, eds. Proceedings of the International Symposium on Raptor Reintroduction, 1985. Arcata, California: Institute for Wildlife Studies.

Oliver, T. C. 1964. Breeding of the LooChoo or Lidth's Jay *Lalocitta lidthi* (Bonaparte). Avic. Mag. 70:212.

Olson, S. L., and H. F. James. 1982a. Fossil birds of the Hawaiian Islands: Evidence for wholesale extinction by man before western contact. Science 217:633-635.

Olson, S. L., and H. F. James. 1982b. Prodromes of the fossil avifauna of the Hawaiian Islands. Smithson. Contrib. Zool. 365.

Olson, S. L., and H. F. James. 1991. Descriptions of thirty-two new species of birds from the Hawaiian Islands. Part 1. Non-passeriformes. Ornithological Monograph No. 46

Owen, D. F. 1959. The breeding season and clutch-size of the Rook *Corvus frugilegus*. Ibis 101:235-239.

Partridge, W. R., 1966. The breeding of the California Scrub Jay (*Aphelocoma coerulescens californica*). Avic. Mag. 72:76-77.

Perkins, R. C. L. 1893. Notes on collecting in Kona, Hawaii. Ibis 5:101-114.

Perkins, R. C. L. 1903. Fauna Hawaiiensis or the Zoology of the Sandwich (Hawaiian) Isles. Vol. 1, Part 4. Cambridge: The University Press.

Picozzi, N. 1975. A study of the carrion/hooded crow in northeast Scotland. Brit. Birds 68:409-419.

Pimm, S. L. 1991. The Balance of Nature? Ecological Issues in the Conservation of Species and Communities. Chicago: University of Chicago Press, 434 pp.

Pimm, S. L., J. M. Diamond, T. Reed, and J. Verner. submitted.

Quammen, D. 1991. A future as big as Indonesia. Outside (July): 33-37.

REFERENCES

Ralls, K., and J. Ballou. 1983. Extinction: Lessons from zoos. Pp. 164-184 in C.M. Schoenwald-Cox, S.M. Chambers, B. MacBryde, and W. L. Thomas, eds. Genetics and Conservation: A Reference for Managing Wild Animals and Plant Populations. Menlo Park, California: Benjamin/Cummings.

Ralls, K. K. Brugger, and J. Ballou. 1980a. Inbreeding and juvenile mortality in small populations of ungulates. Science 206:1101-1103.

Ralls, K. K. Brugger, and A. Glick. 1980b. Deleterious effects of inbreeding in a herd of Dorcas gazelle *Gazella dorcas*. Int. Zoo Yearb. 20:137-146.

Ralls, K., P. H. Harvey, and A. M. Lyles. 1986. Inbreeding in natural populations of birds and mammals. Pp. 35-56 in M.E. Soulé, ed. Conservation Biology: The Science of Scarcity and Diversity. Sunderland, Mass: Sinauer Assocs.

Ralph, C. J., and C. van Riper, III. 1985. Historical and current factors affecting Hawaiian native birds. Bird Conserv. 2:7-42.

Reed, C., and D. Merton. 1991. Behavioral manipulation of endangered New Zealand birds as an aid toward species recovery. Acta XX Cong. Intern. Ornithol. CI: 2514-2522.

Reid, B., and C. Roderick. 1973. New Zealand Scaup *Aythya novaeseelandiae* and Brown teal *Anas auklandica chlorotis* in captivity. Int. Zoo Yearb. 13:12-15.

Richards, P. R. 1973. Breeding the Rook (*Corvus frugilegus*) in captivity. Avic. Mag. 79:82:2-4.

Richards, P. R. 1976. The nesting and handrearing of the Red-billed Blue Magpie *Urocissa erythrorhyncha*. Avic. Mag. 82:2-4.

Richardson, H., N. A. M. Verbeek, and R.W. Butler. 1985. Breeding success and the question of clutch size of Northwestern Crows *Corvus caurinus*. Ibis 127:174-183.

Richner, H. 1992. The effect of extra food on fitness in breeding carrion crows. Ecology 73(1): 330-335.

Risdon, D. H. S. 1960. The breeding of Steller's Jay (*Cyanocitta stelleri*) at Dudley Zoo. Avic. Mag. 66:206-208.

Roles, D. G. 1971. The breeding Mexican Green Jay at the Jersey Zoological Park *Cyanoeorox-yncas*. Avic. Mag. 77:20-22.

Romanoff, A. L. 1972. Pathogenesis of the Avian Embryo: An Analysis of Causes of Malformations and Prenatal Death. New York: John Wiley & Sons. 476 pp.

Roots, C. 1970. Breeding the Southern Tree Pie at the Winged World. Avic. Mag. 76:144-145.

Rowley, I. 1973. The comparative ecology of Australian corvids. IV. Nesting and the rearing of young to independence. CSIRO Wildl. Res. 18:91-129.

Sakai, H. F., and J. R. Carpenter. 1990. The variety and nutritional value of foods consumed by Hawaiian Crow nestlings, an endangered species. Condor 92:220-228.

Sakai, H. F., and C. D. Jenkins. 1983. Capturing the endangered Hawaiian Crow with mist nets. North Am. Bird Band. 8:54-55.

Sakai, H. F., and C. J. Ralph. 1980. Observations on the Hawaiian Crow in South Kona, Hawaii. 'Elepaio 40:133-138.

Sakai, H. F., C. J. Ralph, and C.D. Jenkins. 1986. Foraging ecology of the Hawaiian Crow *Corvus-Hawaiiensis*, an endangered generalist. Condor 88:211-219.

Schoenwald-Cox, C. M., S. M. Chambers, B. MacBryde, and W. L. Thomas. 1983. Genetics and Conservation: A Reference for Managing Wild Animal and Plant Populations. Menlo Park, California: The Benjamin/Cummings. 722 pp.

Scott, J. M., and C. B. Kepler. 1985. Distribution and abundance of Hawaiian native birds: A status report. Bird Conserv. 2:43-70.

Scott, T. A., and M. L. Morrison. 1990. Natural history and management of the San Clemente Loggerhead Shrike. Proc. Western Found Vert. Zool. 4(2):23-60.

Scott, J. M., S. Mountainspring, F. L. Ramsey, and C. B. Kepler. 1986. Forest Bird Communities of the Hawaiian Islands: Their Dynamics, Ecology, and Conservation. Studies in Avian Biology No. 9. Lawrence, Kansas: Allen Press, Inc.

Seal, U. S. 1988. Intensive technology in the care of *ex situ* populations of vanishing species. Pp. 289-295 in E.O. Wilson, ed. Biodiversity. Washington, D.C.: National Academy Press.

Shelton, L. C. 1989. A corvid in your zoo's future? AAZPA Ann. Conf. Proc. 1989:579-587.

REFERENCES

Sherrod, S. K., W. R. Heinrich, W. A. Burnham, J. H. Barclay, and T. J. Cade. 1981. Hacking: A Method for Releasing Peregrine Falcons and Other Birds of Prey. Fort Collins, Colorado: The Peregrine Fund.

Shoffner, R. N. 1948. The reaction of fowl to inbreeding. Poult. Sci. 27:448-452.

Simberloff, D. 1988. The contribution of population and community biology to conservation science. Annu. Rev. Ecol. Syst. 19:473-511.

Sittmann, K., H. Alplanalp, and R. A. Fraser. 1966. Inbreeding depression in Japanese Quail. Genetics 54:371-379.

Skead, C. J. 1952. A study of the Black Crow *Corvus capensis*. Ibis 94:434-451.

Slagsvold, T., J. Sandvik, G. Rofstad, O. Lorentsen, and M. Husby. 1984. On the adaptive value of intraclutch egg-size variation in birds. Auk 101:685-697.

Snyder, N. F. R. 1986. California condor recovery program. Raptor Res. Rep. 5:56-71.

Snyder, N. F. R., and E. V. Johnson. 1985. Photographic censusing of the 1982-1983 California condor population. Condor 87:1-3.

Snyder, N. F. R., and T. Johnson. 1988. Reintroduction of Thick-billed Parrots in Arizona. Pp. 431-436 in B. L. Dresser, R. W. Reece, and E. J. Maruska, eds. 5th World Conference on Breeding Endangered Birds in Captivity. Cincinnati: Ohio, Cincinnati Zoo.

Snyder, N. F. R., and H. A. Snyder. 1989. Biology and conservation of the California Condor. Curr. Ornithol. 6:175-267.

Snyder, N. F. R., and M. P. Wallace. 1987. Reintroduction of the Thick-billed Parrot in Arizona. Pp. 360-384 in A. C. Risser, ed. Second Jean Delacour/IFCB Symposium on Breeding Birds in Captivity. North Hollywood, California: Internat. Foundation for the Conservation of Birds.

Snyder, N. F. R., J. W. Wiley, and C. B. Kepler. 1987. The parrots of Luquillo: Natural history and conservation of the Puerto Rican Parrot. Los Angeles, California: Western Foundation of Bertebrate Zoology.

Springer, P. F., G. V. Byrd, and D. W. Woolington. 1977. Reestablishing Aleutian Canada Geese. Pp. 331-338 in S. A. Temple, ed. Endangered Birds: Management Techniques for Preserving Threatened Species. Madison: University of Wisconsin Press.

Soulé, M. E. 1980. Thresholds for survival: Maintaining fitness and evolutionary potential. Pp. 151-168 in M. E. Soulé and B. A. Wilcox, eds. Conservation Biology: An Evolutionary-Ecological Perspective. Sunderland, Massachusetts: Sinauer Assocs.

Soulé, M. E., ed. 1987. Viable Populations for Conservation. New York: Cambridge University Press, 189 pp.

Soulé, M. E., M. Gilpin, W. Conway, and T. J. Foose. 1986. The millenium ark: How long a voyage, how many staterooms, how many passengers? Zoo Biol. 5:101-113.

Starfield, A. M., and A. M. Herr. 1991. A response to Maguire. Conserv. Biol. 5:435.

Steadman, D. W. 1989. Extinction of birds in Eastern Polynesia: A review of the record, and comparisons with other Pacific island groups. J. Archeol. Sci. 16:177-205

Stiehl, R. B. 1985. Brood chronology of the Common Raven. Wilson Bull. 97:78-87.

Stone, C. P., and L. L. Loope. 1987. Reducing negative effects of introduced animals on native biotas in Hawaii: What is being done, what needs doing, and the role of National Parks. Envrion. Conserv. 14:245-258.

Taylor, J. J. 1985. Iron accumulation in avian species in captivity. Dodo 21:126-131.

Taylor, R. H. 1985. Status, habits and conservation of *Cyanoramphus* parakeets in the New Zealand region. Pp. 195-211 in P. J. Moors, ed. Conservation of Island Birds. Cambridge: International Council for Bird Preservation. Tech. Publ. No. 3.

Temple, S. A. 1977a. The ecology and conservation of kestrels on islands of the Indian Ocean. Pp. 74-82 in R. D. Chancellor, ed. Proceedings of 1975 World Conference of Birds of Prey. Cambridge: International Council for Bird Preservation.

Temple, S. A., ed. 1977b. Endangered Birds: Management Techniques for Preserving Threatened Species. Madison: University of Wisconsin Press.

Temple, S. A., and D. A. Jenkins. 1981. Final Progress Report: 1979 and 1980 Alala Research. 8 pp.

Templeton, A. R., and B. Read. 1983. The elimination of inbreeding depression in a captive herd of Speke's gazelle. Pp. 241-261 in C. M. Schoenwald-Cox, S. M. Chambers, B. MacBryde, and W. L. Thomas, eds. Genetics and Conservation: A Reference for Managing Wild Animal and Plant Populations. Menlo Park, Calif: Benjamin/Cummings.

REFERENCES

Tenovuo, R. 1963. Zur brutzeitlichen Biologie der Nebelkrahe *(Corvus corone cornix* L.) im ausseren Scharenhof Sudwest-finnlands. Ann. Zool. Soc. Vanamo 25:1-147.

Terborgh, J. W., and R. Winter. 1980. Some causes of extinction. Pp. 119-133 in M. E. Soulé and B.A. Wilcox, eds. Conservation Biology: An Evolutionary-Ecological Perspective. Sunderland, Massachusetts: Sinauer Assocs.

Terrasse, M. 1982. Le retour des boulgras. Courr. Nat. 79:15-24.

Thorne, E. T., and E. S. Williams. 1988. Disease and endangered species: The black-footed ferret as a recent example. Conserv. Biol. 2:66-74.

Todd, W. 1980. Breeding Beechey's Jays *Cissilopha beecheii* at the Houston Zoological Gardens. Avic. Mag. 86:123-127.

Todd, D. M. 1984. The release of Pink Pigeons *Columba (Nesoenas) mayeri* at Pamplemousses, Mauritius: A progress report. Dodo 21:43-57.

Tomich, P. Q. 1969. Mammals in Hawaii. Honolulu: Bishop Museum Press.

Tomich, P. Q. 1971. Notes on the nests and behaviour of the Hawaiian Crow. Pac. Sci. 25:465-474.

Toone, W. D., and A. C. Risser, Jr. 1988. Captive management of the California Condor (*Gymnogyps californianus).* Int. Zoo Yb. 27: 50-58.

Triggs, S. J., R. G. Powlesland, and D. H. Daugherty. 1989. Genetic variation and conservation of the Kakapo (*Strigops habroptilus*: Psittaciformes). Conserv. Biol. 3:92-96.

U.S. Fish and Wildlife Service. 1991. Endangered and Threatened Wildlife and Plants. Title 50 CFR 17.11, and 17.12, July 15, 1991. Washington, D.C.: U.S. Fish and Wildlife Service, 37 pp.

Usnik, R. L. 1990. Year end report and recommendations for the Olinda endangered species propagation facility. Maui, Hawaii, 1990. Contract Activities and Final Report, Contract Number: 27460. Unpubl. Report to Hawaii Division of Forestry and Wildlife. 23 pp.

Van Balen, S., and M. N. Soetawidjaya. 1991. Bali Starling Project, Progress Report January-March 1991. International Council for Bird Preservation. unpubl. rept. 6 pp.

Van Balen, S., B. Van Helvoort, and M. N. Soetawidjaya. 1990. Bali Starling Project III, Progress Report July 1987 - September 1990. International Council for Bird Preservation. unpubl. rept., 31 pp.

Van Noordwijk, A. J., and W. Scharloo. 1981. Inbreeding in an island population of the Great Tit. Evolution 35:674-688.

Van Riper, C., III, and S. G. van Riper. 1980. A necropsy procedure for sampling disease in wild birds. Condor 82:85-98.

Van Riper, C., III, S. G. van Riper, M. L. Goff, and M. Laird. 1986. The epizoology and ecological significance of malaria in Hawaiian land birds. Ecol. Monogr. 56:327-344.

Vargas, H. 1987. Frequency and effect of pox-like lesions in Galapagos Mockingbirds. J. Field Ornithol. 58:101-102.

Von Essen, L. 1982. An effort to reintroduce the Lesser White-fronted Goose (*Ansererythropus*) into the Scandinavian mountains. Aquila 89:103-107.

Von Frankenberg, O., E. Herrlinger, and W. Bergerhausen. 1984. Reintroduction of the European Eagle Owl *Bubo b. bubo* in the Federal Republic of Germany. Int. Zoo Yearb. 23:95-100.

Wallace, M. P. 1990. The California Condor: Current efforts for its recovery. End. Species Update 8:32-35.

Wallace M. P. 1991. Methods and strategies for the release of the California Condors to the wild. AAZPA Ann. Conf. Proc. 1991:121-128.

Wallace, M. P., and S. A. Temple. 1987. Releasing captive-reared Andean condors to the wild. J. Wildl. Manage. 51:541-550.

Warner, R. E. 1968. The role of introduced diseases in the extinction of the endemic Hawaiian avifauna. Condor 70:101-120.

Watson, J. 1989. Successful translocation of the endemic Seychelles Kestrel (*Falco araea)* to Praslin. Pp. 363-366 in Meyburg, B. - U. and R. D. Chancellor, eds. Raptors in the Modern World. London: WWGPG

Wayre, P. 1970a. Breeding the Azure-winged Magpie at the Norfolk Wildlife Park, *Cyanopica-cyanus-cooki*. Avic. Mag. 76:240.

REFERENCES

Wayre, P. 1970b. Breeding the Alpine Chough at the Norfolk Wildlife Park, *Pyrrhocorax-graculus-graculus*. Avic. Mag. 76:230-231.

Webber, T., and J. A. Cox. 1987. Breeding and behaviour of Scrub Jays *Aphelocoma coerulescens* in captivity. Avic. Mag. 93:6-14.

Wilcox, B. A. 1980. Insular ecology and conservation. Pp. 95-117 in M. E. Soulé and B.A. Wilcox, eds. Conservation Biology: An Evolutionary-Ecological Perspective. Sunderland, Mass: Sinauer Assocs.

Wiley, J. W., N. F. R. Snyder, and R. Gnam. 1992. Reintroduction as conservation strategy for parrots. Pp. 165-200 in S. R. Beissinger and N. F. R. Snyder, eds. New World Parrots in Crisis: Solutions From Conservation Biology. Washington, D.C.: Smithsonian Institution Press.

Williams, G. R., and D. R. Given. 1981. The red data book of New Zealand. Wellington: Nature Conservation Council.

Wilson, S. B., and A. H. Evans. 1893. Aves Hawaiiensis: The Birds of the Sandwich Islands, Part IV. London: R. H. Porter.

Wingate, D. B. 1982. Successful re-introduction of the Yellow-crowned night heron as a nesting resident on Bermuda. Colonial Waterbirds 5:104-115.

Wingate, D. B. 1985. The restoration of Nonsuch Island as a living museum of Bermuda's pre-colonial terrestrial biome. Pp. 225-238 in P. J. Moors, ed. Conservation of Island Birds: Case Studies for the Management of Threatened Island Species. ICBP Tech. Publ. No. 3. Cambridge: International Council for Bird Preservation.

Wise, C. 1991. The midnight parrot. Exotic Bird Report. 3:3,6.

Witteman, G. J., R. E. Beck, Jr., S. L. Pimm, and S. R. Derrickson. 1990. The decline and restoration of the Guam Rail, *Rallus owstoni*. End. Species Update 8:36-39.

Wittenberg, J. 1968. Freilanduntersuchungen zu Brutbiologie und Verhalten der Rabenkrahe (*Corvus c. corone*). Zool. Jb. Syst. 95:16-146.

Wolinski, Z. 1989. The breeding behaviour of Ravens *Corvus corax* at the Warsaw Zoo. Int. Zoo. Yearb. 28:250-252.

Woolfenden, G. E. 1973. Nesting and survival in a population of Florida Scrub Jays. Living Bird 12:25-49.

Woolfenden, G. E. 1978. Growth and survival of young Florida Scrub Jays. Wilson Bull. 90:1-18.

Woolfenden, G. E., and J. W. Fitzpatrick. 1984. The Florida Scrub Jay: Demography of a Cooperative-Breeding Bird. Princeton, New Jersey: Princeton University Press.

Woolfenden, G. E., and J. W. Fitzpatrick. 1990. Florida Scrub Jays: a synopsis after 18 years of study. Pp. 240-266 in Stacey, P. B. and W. D. Koenig. eds. Cooperative Breeding in Birds: Long-Term Studies of Ecology and Behavior. Cambridge: Cambridge University Press.

Woolfenden, G. E., and J. W. Fitzpatrick. 1991. Florida Scrub Jay ecology and conservation. Pp. 542-565 in Perrins, C. M., J. D. Leberton, and G. J. M. Hirons, eds. Bird Population Studies: Relevance to Conservation and Management. Oxford: Oxford University Press.

Yamamoto, J. T., K. M. Shields, J. R. Millam, T. E. Roudybush, and C. R. Grau. 1989. Reproductive activity of force-paired Cockatiels (*Nymphicus hollandicus*). Auk 106:86 93.

Yom-Tov, Y. 1974. The effect of food and predation on breeding density and success, clutch size and laying date of the Crow (*Corvus corone* L.). J. Anim. Ecol. 43:479-498.

Zimmermann, D. 1951. Zur Brutbiologie der Dohle, *Coloeus monedula* (L.). Ornithol. Beob. 48:73-111.

Zwank, P. J., and S. R. Derrickson. 1981. Gentle release of captive parent-reared Sandhill Cranes to the wild. Pp. 112-116 in J. C. Lewis, ed. Proceedings 1981 Crane Workshop. Crane Workshop [in Grand Teton National Park, Wyo.] Tavernier, Florida: Natl. Audubon Society. Aug. 25-27, 1981.

Zwank, P. J., and C. D. Wilson. 1988. Survival of captive, parent-reared Mississippi Sandhill Cranes released on a refuge. Conserv. Biol. 1:165-168.

Zwank, P. J., J. P. Geaghan, and D. A. Dewhurst. 1988. Foraging differences between native and released Mississippi Sandhill Cranes: Implications for conservation. Conserv. Biol. 2:386-390.

APPENDIX A

The most complete compilations of times to extinction for birds are counts of nesting birds in the annual reports of bird observatories on small islands around Great Britain and Ireland and on Helgoland (an island north of Germany, east of Denmark.) (Pimm et al., submitted). The original sources are mainly the notes of amateur bird-watchers who visited the islands. Those records were summarized and provide estimates of time to extinction of species ecologically and taxonomically similar to the 'Alala.

The islands included are Bardsey, Calf of Man, Copeland, Fair Isle, the Farne Islands, Handa, Helgoland, Hilbre, Isle of May, Lundy, Skokholm, Skolt Head, Skomer, and Steep Holm. These islands were chosen from a much larger set, because their birds had been counted for at least 25 years. Data on five corvid species were examined: Corvus corone (Carrion/Hooded Crows), C. monedula (Jackdaw), C. corax (Raven), Pyrhocorax pyrhocorax (Chough) and Pica pica (Magpie). The selected species are the only corvids on those islands.

Table 1 shows three counts of the number of breeding pairs. In the first example, the species is present at the beginning of the census, present at the end, and does not become extinct. The length of the census, 12 years in this case, is a minimum estimate of the duration that the species lasted. In the second example, the species does become extinct, although again the estimate (5 years) is a minimum, because it is not known when the species first nested on the island. The species invades again, generating a short second record that lacks an extinction. In the third example, a time to extinction is determined. In this case, there are records of when the species was first observed to have nested and when it last nested. This final example shows two complete records for the species. Each record is treated as an independent event. If a species were present in only odd-numbered years, one might suspect that birds were nesting on the island one year, on the adjacent mainland the next. The records would then not be independent. This possibility cannot be completely excluded, but it is probably not a major source of error. The intervals between colonizations were typically much longer than the times to extinction for most of the short-lived populations.

From data like those in Table 1, two statistics can be extracted. The first is a measure of population size. Models of extinction assume that a population ceiling (perhaps imposed by limited suitable habitat), rather than the average population size, sets times to extinction. Reserve managers also encounter such ceilings. The area set aside in a reserve, combined with a characteristic territorial size, imposes a limit on the population that the species is unlikely to exceed, except briefly. Consequently, each example in Table 1 records the maximum number of breeding pairs noted in each year--seven in the first example, five in the second, and two and one for the two records in the third example. There cannot be different true ceilings in the third example, but more than one estimate can be determined.

The second statistic is how long a population lasted from some beginning point until it became extinct. Theory and practice suggest that models measure how long a population lasted from when it was the most abundant. This value is 2 years in the second example and 3 years and 1 year in the two records of the third example in Table 1. Theoretical models often predict times to extinction from a population of a given size. In practice, managers encounter populations likely to be at or close to their maximum size that soon become isolated by habitat fragmentation. A manager's concern is also to predict the times to extinction of a species when it is at or near its population ceiling.

Figure 1 shows the times to extinction from population maximums for complete records (like those in the third example of Table 1) (squares), and times from observed maximums for records where the date of extinction is known, but not the date of colonization (triangles). Also shown are the lengths of the counts when species did not become extinct (circles). All records with maximums less than or equal to 15 pairs are included. The highest maximum count for a population that became extinct on an island was 15 pairs. The median (that is, the fiftieth percentile) times to extinction for population sizes of 1, 2, and 3 pairs is shown in Figure 1. The means were not calculated.[1] The results can be summarized as follows: The median time to extinction for a maximum density of three breeding pairs is 8 years.

Table A.1 Hypothetical counts of corvid species on small islands

Example	No. Breeding Pairs											
	1980	1981	1982	1983	1984	1985	1986	1987	1988	1989	1990	1991
1	1	3	4	2	3	5	6	7	2	3	1	1
2	1	2	3	5	2	0	0	0	0	1	2	2
3	0	1	2	2	1	0	0	0	1	0	0	0

[1] The means have not been calculated for several reasons: Theory suggests and experience confirms that times to extinction are likely to have a highly skewed distribution. It makes little sense to report that populations with maximums of n breeding pairs have a mean time to extinction of, for example, 20 years, when 90% of the populations are extinct in less than 5 years and 1% last for centuries. Criteria for managing species typically involve percentiles. Statements like "under plan A, x% of the populations will be extinct in t years" are the professional standard (Soulé, 1987). Moreover, it is often impossible to calculate a mean, because some of the times to extinction are known to be underestimates.

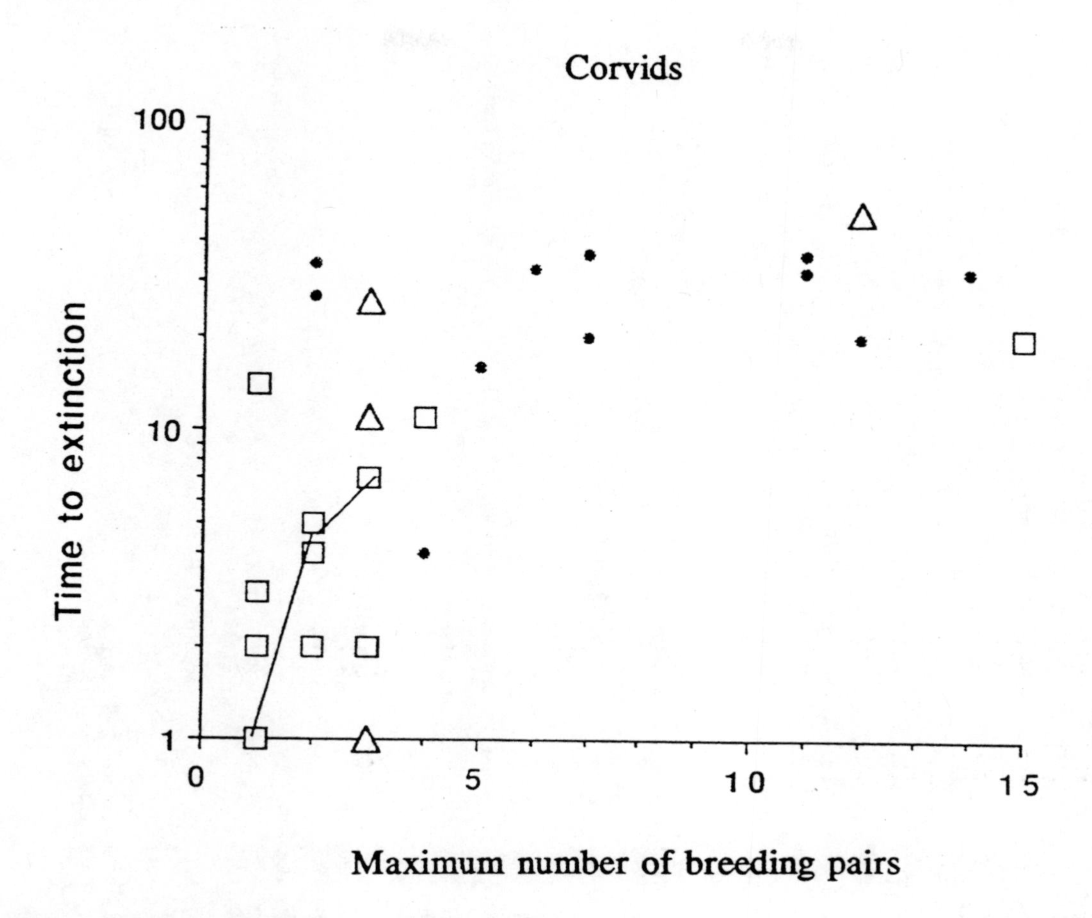

Figure A.1 Times to extinction of corvids on 14 islands off northwest Europe. Line shows median values for each value of observed maximum population size; thus 50% of populations that do not exceed three pairs become extinct in 8 years. Squares, complete records are available (from first invasion to extinction); triangles, only extinction is recorded; circles, minimum estimates where extinction is not recorded.

APPENDIX B

ALALA - HAWAIIAN CROW STUDBOOK

Stud	Sex	Hatch Date	Sire	Dam	Location	Date	Local ID	Event	Birth-Origin	Death-Date	Comments
01	?	~ 1965	Wild	Wild	Hawaii	~ 1965	Unk	Capture	Wild Born		
					Honolulu	~ 1969		Death		~ 1969	
02	M	????	Wild	Wild	Hawaii	????	Wild1	Hatch	Wild Born		Hypothetical Bird
						~ 1978		Death		~ 1978	
03	F	????	Wild	Wild	Hawaii	????	Wild2	Hatch	Wild Born		Hypothetical Bird
						~ 1978		Death		~ 1978	
04	F	????	2	3	Hawaii	????	Wild3	Hatch	Wild Born		Hypothetical Bird
						????		Death		????	
05	M	~ May 1970	Wild	Wild	Hualalai	~ May 1970	Mauka	Hatch	Wild Born		Pericarditis, Splenitis
					Patuxent	~ Jun 1970	Mauka	Transfer			
						10 Oct 1973		Death		10 Oct 1973	
06	F	~ Jun 1970	Wild	Wild	Hualalai	~ Jun 1970	Makai	Capture	Wild Born		Shipment Stress
					Patuxent	~ Jun 1970	Makai	Transfer			
						4 Aug 1970		Death		4 Aug 1970	
07	F	????	Wild	Wild	Honaunau	1 Jun 1973	Hina	Capture	Wild Born		Toxemia
					Patuxent	1 Jun 1973	Hina	Transfer			
					PESPF*	1 Mar 1976	Hina	Transfer			
						7 Aug 1978		Death		7 Aug 1978	
08	M	~ 1 Jun 1973	Wild	Wild	Hualalai	1 Jun 1973	Kekau	Capture	Wild Born		Hemochromatosis
					Patuxent	1 Jun 1973	Kekau	Transfer			
					PESPF	1 Mar 1976	Kekau	Transfer			
						7 Aug 1978		Death		7 Aug 1978	
09	F	~ 8 Jun 1978	Wild	Wild	Hualalai	8 Jun 1978	Hiialo	Capture	Wild Born		Yolk Peritonitis
					PESPF	8 Jun 1978	Hiialo	Transfer			
					OESPF**	1 Nov 1986	Hiialo	Transfer			
						11 Jun 1987		Death		11 Jun 1987	
10	F	~15 Jun 1977	Wild	Wild	Hualalai	15 Jun 1977	Iola-E	Capture			Hemochromatosis
					PESPF	15 Jun 1977	Iola-E	Tranfer			
						21 Jun 1979		Death		21 Jun 1979	
11	M	~ 1 May 1978	Wild	Wild	Hualalai	15 Jun 1978	Ulu	Capture	Wild Born		
					PESPF	15 Jun 1978	Ulu	Transfer			
						2 Jul 1981		Death		2 Jul 1981	
12	M	~ 1 May 1978	Wild	Wild	McCandless	21 Jun 1978	Imia	Capture	Wild Born		Hemochromatosis
					PESPF	21 Jun 1978	Imia	Transfer			
						1 Mar 1981		Death		1 Mar 1981	

Stud	Sex	Hatch Date	Sire	Dam	Location	Date	Local ID	Event	Birth-Origin	Death-Date	Comments
13	M	~ 1 Jun 1973	Wild	Wild	Yee Hop	1 Jun 1973	Umi	Capture	Wild Born		
					Patuxent	1 Jun 1973	Umi	Transfer			
					PESPF	1 Mar 1976	Umi	Transfer			
					OESPF	1 Nov 1986	Umi	Transfer			
14	F	~ 1 May 1977	2	3	Hualalai	28 Jun 1977	Luukia	Capture	Wild Born		
					PESPF	28 Jun 1977	Luukia	Transfer			
					OESPF	1 Nov 1986	Luukia	Transfer			
15	F	~ 1 May 1978	Wild	Wild	Honaunau	19 Jun 1978	Mana	Capture	Wild Born		Yolk Peritonitis
					PESPF	19 Jun 1978	Mana	Transfer			
					OESPF	1 Nov 1986	Mana	Transfer			
						28 Jul 1991		Death		28 Jul 1991	
16	F	~ 1 May 1981	Wild	Wild	Honaunau	24 Jun 1981	Kolohe	Capture	Wild Born		Imprinted
					PESPF	24 Jun 1981	Kolohe	Transfer			
					OESPF	1 Nov 1986	Kolohe	Transfer			
17	M	~16 May 1981	13	14	Honolulu	16 May 1981	Kelii	Hatch	Captive Born		
					PESPF	1 Aug 1981	Kelii	Transfer			
					OESPF	1 Nov 1986	Kelii	Transfer			
18	M	~20 May 1981	13	14	Honolulu	20 May 1981	Keawe	Hatch	Captive Born		
					PESPF	1 Aug 1981	Keawe	Transfer			
					OESPF	1 Nov 1986	Keawe	Transfer			
19	M	~18 May 1981	13	14	Honolulu	18 May 1981	Kalani	Hatch	Captive Born		
					PESPF	1 Aug 1981	Kalani	Transfer			
					OESPF	1 Nov 1986	Kalani	Transfer			
20	F	~ 1 May 1983	Wild	4	Hualalai	20 Jun 1983	Waalani	Capture	Wild Born		
					PESPF	20 Jun 1983	Waalani	Transfer			
					OESPF	1 Nov 1986	Waalani	Transfer			
21	F	~11 Jun 1988	19	14	OESPF	11 Jun 1988	Hooku	Hatch	Captive Born		
22	M	~24 May 1989	19	14	OESPF	24 May 1989	Hoikei	Hatch	Captive Born		
23	M	~23 May 1990	18	15	OESPF	23 May 1990	Kinohi	Hatch	Captive Born		

* PESPF Pohakola Endangered Species Propagation Facility
** OESPF Olinda Endangered Species Propagation Facility

Notes:

1. Totals: 11 males, 11 females, 1 unknown (23)
2. Data complied by Cynthia Kuehler, Zoological Society of San Diego

BIOGRAPHICAL STATEMENTS

W. Donald Duckworth is president and director of the Bishop Museum in Honolulu, Hawaii. Trained in systematic entomology, his research has focused on lepidoptera in the tropics.

Tom J. Cade is professor emeritus in the Section of Ecology and Systematics, Cornell University and professor and director of raptor research at Boise State University, Idaho. He is the founding chairman of The Peregrine Fund, Inc. Much of his research has dealt with falcons and other birds of prey.

Hampton L. Carson is professor emeritus in the Department of Genetics and Molecular Biology, John A. Burns School of Medicine, University of Hawai'i at Manoa, Honolulu. His research career has been devoted to field and laboratory analyses of the genetic changes that occur in natural populations of both continental and insular species.

Scott R. Derrickson is curator of ornithology and deputy associate director for conservation of the Smithsonian Institution's National Zoological park. His research has focused primarily on endangered-species propagation and reintroduction.

John W. Fitzpatrick is executive director and senior research biologist at Archbold Biological Station, Lake Placid, Florida. His research includes ecology and population biology of the Florida Scrub Jay, community ecology of tropical forest birds, and the role of cattle ranches as wildlife habitat.

Frances C. James is in the Department of Biological Science at Florida State University. Her research emphasizes avian ecology, biogeography, and conservation. She is a past member of the Board of Directors of the World Wildlife Fund and president of the American Ornithologists' Union.

Special Advisors

Cynthia Kuehler is the curator of zoology for the San Diego Zoo and San Diego Wild Animal Park. Her research deals with developing techniques for propagation of endangered animals in captivity.

Stuart Pimm is in the Department of Zoology and Graduate Program in Ecology at the University of Tennessee at Knoxville. His research deals with issues of conservation biology.